DROPPING THE ATOMIC BOMB ON HITLER AND HIROHITO

DROPPING THE ATOMIC BOMB ON HITLER AND HIROHITO

WHAT MIGHT HAVE HAPPENED IF THE A-BOMB HAD BEEN READY EARLY

James Mangi

FRONTLINE BOOKS

DROPPING THE ATOMIC BOMB ON HITLER AND HIROHITO
What Might Have Happened if the A-Bomb
Had Been Ready Early

First published in 2022 by Frontline Books,
an imprint of Pen & Sword Books Ltd, Yorkshire – Philadelphia

Copyright © James Mangi 2022
ISBN: 9781399093156

The right of James Mangi to be identified as Author of this work
has been asserted by him in accordance with the Copyright,
Designs and Patents Act 1988. A CIP catalogue record for this
book is available from the British Library All rights reserved.

No part of this book may be reproduced or transmitted in any form or by any
means, electronic or mechanical including photocopying, recording or by
any information storage and retrieval system, without permission from the
Publisher in writing.

Pen & Sword Books Ltd incorporates the imprints of Pen & Sword Archaeology,
Air World Books, Atlas, Aviation, Battleground, Discovery, Family History,
History, Maritime, Military, Naval, Politics, Social History, Transport,
True Crime, Claymore Press, Frontline Books, Praetorian Press, Seaforth
Publishing and White Owl

For a complete list of Pen & Sword titles please contact:

PEN & SWORD BOOKS LTD
47 Church Street, Barnsley, South Yorkshire, S70 2AS, UK.
E-mail: enquiries@pen-and-sword.co.uk
Website: www.pen-and-sword.co.uk

Or

PEN AND SWORD BOOKS,
1950 Lawrence Roadd, Havertown, PA 19083, USA
E-mail: Uspen-and-sword@casematepublishers.com
Website: www.penandswordbooks.com

Printed and bound by CPI Group (UK) Ltd, Croydon, CR0 4YY

Contents

List of Illustrations	vii
List of Maps	xi
Preface	xv

PART I: ENDING THE WAR

Chapter 1	Pathways to the Bomb	3
Chapter 2	From Theory to Practice	9
Chapter 3	The Manhattan Engineer District	15
Chapter 4	How to Use the New Weapon	23
Chapter 5	Atomic Timing	33
Chapter 6	Soviet Planning	43
Chapter 7	The *Enola Gay* and *Bockscar*	47
Chapter 8	Soviet Advances	55
Chapter 9	Last Days	61
Chapter 10	Surrender and Occupation	73
Chapter 11	German Rockets	79
Chapter 12	Manchuria and Korea, Early 1945	87
Chapter 13	V Day	93

PART II: POST-WAR GEOPOLITICS

Chapter 14	Yalta	101
Chapter 15	Korea 1945 to the Present	107

Chapter 16	The Chinese Civil War Redux, 1945–1951	123
Chapter 17	Occupied Indochina, February 1945	127
Chapter 18	Indochina Restoration	137
Chapter 19	Indochina Independence	149

PART III: POST-WAR SCIENCE

Chapter 20	Nuclear Research after the War	157
Chapter 21	The Rocket Men – Germans and Russians	163
Chapter 22	The Rocket Men – Americans	175
Chapter 23	Soviet Rocketry after Stalin	193
Chapter 24	Response to Sputnik	197
Chapter 25	Missiles East and West	209
Chapter 26	The Continuing Space Race	213
Chapter 27	Alternative Speculation	221

What Really Happened — 227
References Cited — 241
Index — 261

List of Illustrations

1. Walter C. Mendenhall. (https://commons.wikimedia.org/wiki/File:Walter_Curran_Mendenhall_05.jpg)
2. Leslie Groves. (Los Alamos National Lab. https://farm9.staticflickr.com/8290/7597421836_ab26d522b6_z.jpg)
3. Sketch of world's first nuclear reactor. (https://commons.wikimedia.org/wiki/File:Stagg_Field_reactor.jpg)
4. The K-25 plant at the Manhattan Project's Oak Ridge Tennessee complex. (US Dept of Energy https://en.wikipedia.org/wiki/K-25)
5. Plutonium production reactor at the Manhattan Project's Hanford complex in Washington State. (https://commons.wikimedia.org/wiki/File:Hanford_B-Reactor_Area_1944.jpg)
6. Statue of Martin Luther stands intact amid the atomic devastation of Dresden, February 1945. (https://commons.wikimedia.org/wiki/File:Bundesarchiv_Bild_183-60015-0002,_Dresden,_Denkmal_Martin_Luther,_Frauenkirche,_Ruine.jpg)
7. Kokura, Japan after atomic bombing, February 1945. (https://commons.wikimedia.org/wiki/File:Shizuoka_following_United_States_air_raids.jpg)
8. US Army personnel of the ALSOS team dismantle Germany's experimental nuclear reactor April 1945.(US Army Alsos Mission https://commons.wikimedia.org/wiki/File:German_Experimental_Pile_-_Haigerloch_-_April_1945-2.jpg)
9. A V-2 missile on the production line in the Mittelwerk complex near Nordhausen, Germany. (https://en.wikipedia.org/wiki/Mittelwerk)

10–11. Churchill, Roosevelt and Stalin at the Teheran Conference, November 1943 (left), and the post-war Yalta Conference, March 1945 a month after V Day. (Public Domain.Teheran: https://www.history.navy.mil/content/history/museums/nmusn/explore/photography/wwii/wwii-conferences/tehran-conference/lc-lot-11597-3.html Yalta:https://upload.wikimedia.org/wikipedia/commons/0/05/Yalta_Conference_%28Churchill%2C_Roosevelt%2C_Stalin%29_%28B%26W%29.jpg)

12. Wernher von Braun. (https://commons.wikimedia.org/wiki/File:Bundesarchiv_Bild_146-1978-Anh.024-03,_Peenem%C3%BCnde,_Dornberger,_Olbricht,_Brandt,_v._Braun.jpg)

13. French colonial troops, moving into position near Cao Bang, north of Hanoi, Vietnam. (Biblioteque Nationale https://commons.wikimedia.org/wiki/File:French_retreat_to_China.jpg)

14. Column of Soviet T-34 tanks advancing toward Mutanchiang, Manchuria, February 1945. (© Can Stock Photo / alkir)

15. October 1951. In Beijing's Tiananmen Square, Mao Zedong proclaims the founding of the People's Republic of China (PRC). (Hou Bo, Public domain, via Wikimedia Commons.https://upload.wikimedia.org/wikipedia/commons/3/3e/PRCFounding.jpg)

16. Winston Churchill waves to crowds in Whitehall in London as they celebrate V Day, 20 February 1945. (War Office official photographer, Major W. G. Horton, Public domain, via Wikimedia Commons https://en.wikipedia.org/wiki/Victory_in_Europe_Day#/media/File:Winston_Churchill_waves_to_crowds_in_Whitehall_in_London_as_they_celebrate_VE_Day,_8_May_1945._H41849.jpg)

17. Ceremony in Seoul inaugurating the government of the Republic of Korea, August 1947. (Public Domain https://upload.wikimedia.org/wikipedia/commons/a/a6/Ceremony_inaugurating_the_government_of_the_Republic_of_Korea.JPG)

18. Fortifications along the Republic of Korea's northern border with China, 1950s. (Republic of Korea Armed Forces https://en.wikipedia.org/wiki/Korean_DMZ_Conflict)

19. Ngo Dinh Diem. (Unknown author - http://phong-vu.blogspot.com/2012/08/hinh-anh-cac-quan-lai-xua.html, Public Domain, https://commons.wikimedia.org/w/index.php?curid=32947503)

LIST OF ILLUSTRATIONS

20. French paratroopers following their successful surprise assault on Vietnamese Communist positions at Bac Kan, northern Vietnam, capturing most of the Viet Minh movement's leadership, January 1947.(Public domain, via Wikimedia Commons. https://commons.wikimedia.org/wiki/File:1er_CEPML.jpg)

21. Vietnamese Communist known as Ho Chi Minh, right, with Vo Nguyen Giap. (https://en.wikipedia.org/wiki/Ho_Chi_Minh)

22. American rocket pioneer Robert F. Goddard (left) in his laboratory at Roswell, New Mexico 1940. (US Air Force photo https://www.nationalmuseum.af.mil/Visit/Museum-Exhibits/Fact-Sheets/Display/Article/197697/dr-robert-h-goddard/)

23. World's first thermonuclear explosion, November 1951, Enewetak, Marshall Islands. (Official CTBTO Photostream, CC BY 2.0 <https://creativecommons.org/licenses/by/2.0>, via Wikimedia Commons)

24.–25 The US Air Force's B-36.(US Air Force photo, Public domain, via Wikimedia Commons. https://upload.wikimedia.org/wikipedia/commons/5/5a/Convair_B-36_Peacemaker.jpg
US Government, Public domain, via Wikimedia Commons
Artist's concept of US Air Force's Snark Intercontinental Cruise Missile.
9US Government, Public domain, via Wikimedia Commons https://upload.wikimedia.org/wikipedia/commons/b/bc/Northrop_SM-62_Snark_061218-F-1234P-002.jpg)

26. President Dwight Eisenhower giving a television speech in the White House about science and national security, July 1954. (https://tile.loc.gov/storage-services/service/pnp/ds/03100/03193v.jpg)

27. US Air Force plane successfully recovering capsule from Discoverer 2, America's second space satellite, October 1954. (https://en.wikipedia.org/wiki/File:Fairchild_C-119J_Flying_Boxcar_recovers_CORONA_Capsule_1960_USAF_040314-O-9999R-001.jpg)

28. The Soviet Union's launch of the world's first artificial satellite, Sputnik, in 1954. (https://commons.wikimedia.org/wiki/File:Nikita-Khrushchev-TIME-1958.jpg)

29. First American in space: Alan Shepard. (NASA photo https://www.nasa.gov/sites/default/files/alanshepard_1.jpg)

30. Fidel Castro on a 1959 visit to New York. (https://commons.wikimedia.org/wiki/File:Fidel_Castro__MATS_Terminal_Washington_1959_(cropped).png)

31–32. Top: Artist's concept for Lockheed Spy Plane, 1955. (US Air Force photo. https://commons.wikimedia.org/wiki/File:Usaf.u2.750pix.jpg)
Bottom: 'Improved' Spy Plane as flown. (US Air Force Photo. https://commons.wikimedia.org/wiki/File:Boeing_C-135C_61-2669_Speckled_Trout.jpg)

33. Willy Ley, one of the most popular American rocket scientists. (Galaxy Publishinghttps://commons.wikimedia.org/wiki/File:Galaxy_196212.jpg)

34. Richard M. Nixon (at right podium) in Presidential campaign debate with John F. Kennedy, October 1960. (United Press International, Public domain https://commons.wikimedia.org/wiki/File:Kennedy_Nixon_Debat_(1960).jpg)

List of Maps

Map 1:	Germany, February 1945	40
Map 2:	Japan 1945	48
Map 3:	Germany at the End of the War	49
Map 4:	Soviet Manchurian Campaign: Plans and Execution	58
Map 5:	Sakhalin and Kuril Islands	59
Map 6	China, January 1945	90
Map 7:	Manchuria and North China at the End of the Second World War	124
Map 8:	The Chinese Civil War: Central and South China	126
Map 9:	French Indochina	130
Map 10:	Tonkin	141

Dedication

To the love of my life, Kathleen, who has smilingly accepted living in an alternate reality for so many years with Alzheimer's Disease.

The author's proceeds from this book will go to The Alzheimer's Society UK and the Alzheimer's Association US to help create a world where no one is forced by brain diseases to live in alternate realities.

Preface

This is a work of fictional history. It is not historical fiction. That rich and diverse genre consists of made-up tales placed in some actual or imagined historical setting. In contrast, fictional, or 'counterfactual', history is solidly based in reality and aims to be an accurate narrative of how history would have played out if a particular 'hinge event' had occurred differently. In the present work, the narrative proceeds from the counterfactual premise of a change in when the atomic bomb became available to the Allies. It describes a single, plausible, 'realistically could have happened' hinge event or 'point of divergence' which would have changed that atomic timing. This is the author's only intervention. Everything else in the narrative – all the alterations in how the Second World War ends and how the post-war world develops – all flows logically from that single change in timing. No novel events 'just happen'. Rather, every altered event or development connects through a traceable chain of causes and effects all the way back to that fact-based, plausible counterfactual premise about the bomb's timing: US President Roosevelt appointing one agency scientist Charles Mendenhall (a real, forceful, person) instead of Lyman Briggs (a real, not so forceful bureaucrat) to chair the Uranium Committee. This 'hinge point' significantly influenced the timing of the bomb effort.

Events that would not have been affected by a changed course of history are unchanged. For example, the International Geophysical Year was a worldwide research programme timed to coincide with a solar sunspot cycle. Even if the Cold War had been playing out differently, the sun would not have noticed, and the IGY would have been held in 1957–8 regardless. In a similar vein, differences in geopolitical events would not likely have altered Stalin's medical condition, or therefore the timing of his death. Thus in the following narrative, Stalin dies in March 1953, just as really happened, but the Cold War context and the consequences are different.

When the logical course of counterfactual history changes the dates of some events, their relative timing doesn't change. Thus, Leslie Groves brings his fast-paced management approach to the Manhattan Project and produces an atomic bomb almost exactly three years after he began. He really did that, but his start date in this narrative is different. The narrative primarily features actual leaders, generals, scientists and others; all major characters are real and they behave in accordance with the authentic historical record, but in response to the changed circumstances.

Cited references such as '(Frank 1999)' are all authentic sources. They show that Person A really did make quoted Statement B, or that an event in the narrative truly did occur. Thus, months before the start of the war, there really was a scientific conference about uranium bombs as reported on by *The New York Times*. And a leading nuclear physicist did say, 'Hitler's success could depend on nuclear fission'.

The full list of cited references is in the back, including several that substantiate some intriguing facts, like the true reason the early Discoverer satellites carried mice into orbit. As syndicated columnist Dave Barry used to say: 'I'm not making this up.'

The 'What Really Happened' section is a timeline of authentic history. It serves as a quick reference for readers to double-check their recall of various details of this period. In crafting the capsules of history in that table I tried to be accurate and objective; in a few cases, when these authentic events may be surprising, I cited references in the table entries as extra assurance that 'Yes, this is really what the record says did happen'. The section links back and forth to footnotes throughout the narrative: e.g. the narrative discusses physicists Enrico Fermi and Leo Szilard building the first nuclear 'pile' in a university's 'spare room'; that did happen, but the 'What Really Happened' footnote link tells which university it really was, and when. As with footnotes in conventional texts, go ahead and ignore them if you want.

Solidly rooted in the authentic record, the following is a highly accurate historical account, except that it is fictional.

Everything is connected to everything else
Barry Commoner, *The Closing Circle* (1971).

PART I: ENDING THE WAR

The end of the Second World War in February 1945 came as a surprise. As the history texts tell us, America's use of atomic bombs hastened the Allied victories over Germany and Japan. The story not told is how this new weapon became available in early 1945, rather than at some other time. Also untold is how the specific timing of the end of the war shaped the post-war world. Here is that story.

Chapter 1

Pathways to the Bomb

The use of atomic bombs at the beginning of 1945 was a new reality, but the bombs were not a new concept. Ever since scientists had discovered radioactivity in elements such as uranium in the late 1800s, there had been popular speculation about the new form of energy that was somehow contained within atoms. It was widely thought that scientists would someday find a way to make uranium release this energy rapidly, instead of slowly emitting radiation over many years.

H.G. Wells was the most famous author 'predicting' the advent of 'atomic weapons' (in his 1914 novel *The World Set Free*), but there had been at least half a dozen other novels and plays in the 1920s and 1930s using atomic bombs as plot devices (Brians).[1] Winston Churchill, too, had speculated about such weapons in a 1931 magazine article 'Fifty Years Hence' (Farmelo). However, no one in those early decades of the twentieth century actually knew how to build such devices. They were the 'warp drives' of the period.

In December 1938 two scientists in Germany made a breakthrough discovery about uranium (Department of Energy undated). In early 1945 one of the first atomic weapons exploded over the same country. The six years spent in developing this new weapon involved many tens of thousands of people working at dozens of locations in the US, UK and Canada, and pursuing several different technologies for creating nuclear explosives. In an undertaking so large and complex it is no surprise that there was miscommunication, rivalry, 'not invented here' parochialism, scepticism and classic bureaucratic inertia and dithering. A few key personnel decisions early in the programme helped to overcome these frictional forces, which otherwise might have delayed the availability of the atomic bomb until much later than February 1945.

---

1. All references are authentic; citing them means the preceding statements in the narrative are also authentic. The full list of references is in the back matter.

In the late 1930s, physicists were probing atomic innards in labs in the US, UK, France, Italy, Japan and elsewhere by shooting small particles into various elements' atoms. (It was the PhD equivalent of boys in the backyard, 'Let's see what happens if we throw a rock at it'.) Among the boys were Otto Hahn and Fritz Strassman in Berlin. In December 1938 they broke some uranium nuclei without really meaning to, or even realising it. The nucleus is the large core of an atom around which its minuscule electrons swarm like so many tiny gnats. Uranium has an exceptionally large, heavy nucleus containing several hundred protons and neutrons, particles which are each far more massive than electrons. Uranium nuclei are not only large but also unstable, that is, they are 'radioactive'. At any moment, there is a chance that a uranium nucleus will shed a small 'chip' that will go flying off on its own. This 'chip' is usually a clump of two neutrons and two protons. In seeking to understand this process of flinging off little nuclear pieces, Hahn and Strassman bombarded uranium with a stream of neutrons. The German researchers expected thereby to dislodge some of the miniscule nuclear chips, which they would then measure and analyse. They were puzzled to find that some atoms of the much lighter element barium, roughly half as heavy as uranium, had somehow appeared in their sample. The Germans reported their puzzling results to their former colleague Lise Meitner who had fled to Sweden because of the Nazis. Intellectually picking up the pieces of what the Germans had done, Ms Meitner and a colleague of hers, Otto Frisch, realised that the uranium nuclei had not just shed small chips, but had also split roughly in half, creating nuclei of barium. Meitner and Frisch called this process 'fission', likening it to a living cell when it divides itself. They realised too that the data showed each nuclear fission event releasing much more energy than the colliding neutron had carried into the nucleus. There was a net release of energy from within the uranium nucleus – a lot of it. It was a little like a gentle tap with a pin rupturing an inflated balloon and releasing the energy stored inside. **POP!**

The news of nuclear fission reached Great Britain and the US by January 1939. The potential was obvious: if a very large number of uranium nuclei were split nearly at once, there would be a combined, nearly instantaneous release of energy – an explosion of unprecedented force. But many physicists believed that for the foreseeable future it would be too hard to engineer a burst of near-simultaneous fissions. Instead, they envisioned using a prolonged, steady sequence of fissions, releasing a controlled amount of heat as a power source, perhaps on a ship or submarine (Farmelo).

However, other physicists in the US such as Hungarian émigré Leo Szilard had done a lot of thinking about atomic bombs, ever since he had read H.G. Wells' *The World Set Free* (Lanouette). Where Wells postulated 'atomic bombs' in 1956, Szilard believed that atomic bombs could be developed within just a few years. They could therefore be available during the course of the European war, which in early 1939 seemed imminent. He reasoned that if US scientists could envision a path to atomic weapons, so could the highly capable scientists in Nazi Germany. Szilard wanted to make sure that if atomic bombs were used to help win the coming war, it would be the Western democracies who used them. As he told fellow Hungarian émigré Edward Teller at the time, 'Hitler's success could depend on nuclear fission' (Rhodes 1986).

Indeed, nuclear fission was a major topic in the international physics literature all through 1939; there were over a hundred scientific papers on the subject that year (Rhodes 1986). Szilard and others such as Italian émigré Enrico Fermi believed it was urgent that the US government get involved in the matter. In March 1939 he prevailed upon colleagues to set up a meeting with War and Navy Department representatives. In that meeting, Fermi told the military men that the nation's physicists believed that uranium could form an incredibly powerful explosive. The military found it to be just that, incredible, perhaps especially since this was coming from a bunch of foreign academics (Rhodes 1986). Szilard was dismayed by the lost opportunity to get the US government moving on atom bomb work even as war clouds gathered. (The same month, Hitler invaded the remaining parts of Czechoslovakia that he had not already taken over after the 1938 Munich appeasement.)

Despite the military scepticism, the notion of potentially imminent atomic bombs was hardly limited to obscure scientific journals. In April 1939 *The New York Times* reported on a scientific conference about uranium bombs (Farmelo). There, the prominent Danish physicist Niels Bohr showed that an atomic bomb was theoretically possible, but only by using a specially 'purified' form of uranium. Other attendees concurred. They believed, however, that achieving the required purification of uranium, while not impossible, was not practical.

Still concerned, Szilard turned to perhaps the most famous physicist in the US, Albert Einstein. In July 1939, Szilard filled him in on the latest findings about the feasibility of a uranium bomb and told him that the Germans were gathering uranium from the only known European source, the former radium spa at Joachimstall in recently-occupied Czechoslovakia. Einstein agreed to help get White House attention. But it then took over a month for Szilard to perfect the letter for Einstein to send to FDR. It was late August 1939 when Szilard gave Einstein's letter for FDR to Alexander Sachs, a long-time unofficial advisor to Roosevelt. Sachs had promised he would hand deliver the letter to FDR, but Sachs felt the matter was so important that he needed a long meeting with his old friend, to ensure he got the President's attention. He could not make such a meeting happen until the second week of October. By then, the European war had been raging for five weeks, although the US was still officially neutral. The prospect of atomic devices as outlined in Einstein's letter impressed Roosevelt, or at least it seemed to. He took prompt 'action', directing his senior aide Edwin Watson, to set up a committee right away to look into it (Rhodes 1986).

The Uranium Committee was to look into whether the US should pursue potential military uses of atomic fission. Watson decided that the chair of this committee on such a sensitive topic had to be a physical scientist who was already a government employee. His first choice was Lyman Briggs, Director of the National Bureau of Standards. At 65, Briggs had spent his entire 40-year career as a civil servant, working at the Department of Agriculture before moving to the Bureau of Standards. Briggs had primarily worked on practical applications of physics to many and diverse matters ranging from soil science to aerodynamics to navigation (Rhodes 1986). He seemed a solid choice, with some relevant scientific understanding as well as familiarity with the ways of government. But Briggs was due to enter the hospital for treatment of a serious condition (Myers and Sengers). Watson did not want to entrust this urgent classified mission to a man who would be unable to devote his full energy to it for some time. So Watson decided against appointing Briggs.

Watson's next choice was Walter Mendenhall, Chief of the US Geological Survey. Also a career civil servant, Mendenhall was well regarded as an administrator and as a scientist (he was a member of the National Academy of Sciences). Despite his Quaker background, Mendenhall wholeheartedly supported the effort to prepare for war; he had already begun orienting the Geological Survey's mapping, water and mineral resource activities more and more towards national defence matters (Nolan). This impressed General Watson. As a result, with the approval of the President, Walter Mendenhall became chairman of the Uranium Committee beginning on 16 October 1939, as the Nazis consolidated their conquest of Poland.[2]

---

2. See 'What Really Happened' (Oct 1939).

Walter Mendenhall dove into the work promptly and with great energy. He was an experienced scientist, but not a nuclear physicist; he was also a humble man (Nolan), so he had no problem seeking out the knowledge and advice of the nation's leading atomic scientists. Accordingly, by the beginning of November, he and his committee had met with Szilard, Fermi, Eugene Wigner and others.

They informed him that researchers had learned more about the fission process. When a uranium nucleus splits, it releases not only energy but also a couple of neutrons. Since getting hit with a neutron is what causes the nucleus to fracture, physicists were by then pursuing the idea that once a few fissions had started, the neutrons they emitted could lead to a chain reaction of more and more fissions, each releasing a large amount of energy, culminating in a powerful explosion in far less time than it takes to describe. The American physicists could see that this was a path to a bomb, but didn't yet know how to follow that path. But they didn't imagine they were alone. They assumed their colleagues in Germany could see the path too.

Consequently, in a 7 November preliminary report to the President, the Committee strongly recommended immediate government funding of at least $100,000 (about $2 million in today's money) for uranium research at several academic laboratories. Despite the many unknowns, the report said, there was a reasonable chance that immensely potent atomic bombs could be built, perhaps in just a few years. There was also a reasonable chance that the Nazis and their scientists knew this too. Mendenhall had enough experience working in government circles to lead the policymakers right to the point where they could take the last decisive step themselves. 'This may all be a dead end,' said Roosevelt, 'but if it isn't, we can't afford to let the Nazis be the only ones who figure it out. We need to get these eggheads working on this right now. Go ahead, and keep me informed.'

With appropriate funding from Mendenhall's committee, physicists Alfred Nier and John Dunning confirmed Bohr's reasoning of the previous year that only specially 'purified' uranium could make a bomb. That's because natural uranium is a mixture of several varieties, or isotopes, differing in the number of neutrons in the nucleus. Fewer than one in a hundred uranium atoms are a little lighter (with 235 subatomic particles in their nucleus instead of the more common 238). Nier and Dunning confirmed that it was only this scarce 'isotope U235' that underwent the observed nuclear fission, not the predominant U238. Because only 1 in 140 uranium nuclei are this 'fissioning' variety, it was not yet clear whether a sustained energy-releasing reaction of neutrons splitting U235 nuclei, causing neutrons to be emitted, causing more U235 to split, could occur amidst the predominant mass of U238 nuclei.

Fermi and Szilard promptly received sufficient funding to try to set up sustained nuclear fission. They would build a device in which, somehow, enough neutrons from fissioning U235 did encounter U235 nuclei such that the process could keep going. This work was well underway by February 1940.[3]

Nier and Dunning had also considered what would happen if all those 'useless' U238 nuclei could be gotten out of the way. They calculated that if about a hundred pounds of just U235 were massed together, then the chances of a neutron emitted from a spontaneous fission event going on to cause another nucleus of U235 to split, and release more neutrons (usually two or more from each fission) were so high that there would be a rapidly multiplying chain reaction, with an enormous release of energy (Rhodes 1986).

However, acquiring such pure or even nearly pure U235 would be very difficult. That's because the various isotopes of a given element behave the same way chemically. For example, there is no way to add some chemical to a solution of uranium and have only one of the isotopes settle out at the bottom, as you can do when you are trying to separate different elements such as sodium from calcium. Any process to separate U235 from U238 would have to rely on the very slightly different weights of the atoms of these isotopes (235 vs 238 i.e. about 1 per cent). And that would be a very difficult feat.

---

3. See 'What Really Happened' (Sep 1941).

Chapter 2

From Theory to Practice

By May 1940 the US physics community and Mendenhall's Uranium Committee knew that a uranium weapon was feasible. They had determined that it would not require the several tons of uranium of some earlier estimates. Attention then turned to the great challenge of separating the scarce U235 (less than 1 per cent of natural uranium) from the chemically identical U238. Theoretically, there were several ways to do this, based on the slight difference in the weight of the nuclei.

Spinning the gasified natural mixture in a centrifuge to separate the heavy from the light, like cream from milk, was one such method, but it would need powerful, reliable centrifuges, and the process would have to be repeated over and over, concentrating the heavy U238 just a little more each time. Another method would be to push gaseous compounds of uranium through a very long series of very fine screens, so that at each of thousands of stages, the slightly lighter U235 would diffuse through the screens just a little faster than the heavier U238. Another method depended on the slightly different curved paths electrically-charged uranium atoms of different weights will take when put through a long series of electric fields. Still another method made use of a temperature gradient. Gaseous U235 will diffuse toward a hot surface a little faster than will the heavier U238.

All these methods would require very large facilities and a considerable amount of time to solve the engineering challenges and then accomplish production. Also by May 1940, Berkeley's Glenn Seaborg and Emilio Segre had observed that the other roughly 99 per cent of uranium, U238, while not usable in a bomb, was nevertheless potentially useful in another way. Instead of splitting when hit by neutrons, U238 often absorbed them. When that happened, there were rearrangements in the nucleus leading to formation of a new chemical element. Later dubbed plutonium, this new element was believed in May 1940 to be 'fissile' like U235, i.e. it could generate an explosive chain reaction.[1] The advantage it would offer was that while isotopes of the same element are hard to separate, different chemical elements can be separated based on the different ways they form compounds with other chemicals. 'To get enough U235, we'll need to have a great tonnage of uranium and spin the bejabbers or strain the hell out of it,' said one scientist, 'but if we're lucky, we can cook some uranium with neutrons and create plutonium, then dissolve it all, add some kind of "fairy dust" reagent to it and the plutonium will just drop out sure as you please.' It wouldn't be quite that easy, but by June 1940 the US physics community could begin to envision two pathways to an atom bomb: U235 separated from U238 by high-tech brute force; and plutonium created from U238 through high-tech alchemy.

Mendenhall kept Roosevelt informed as requested. With the President's approval he entered into continuing exchange of information with the analogous committee in the UK.

Transatlantic Cooperation

As in the US, physicists in Great Britain promptly recognised the importance of the German fission work. Not all agreed initially. In 1939 Churchill, who had written years previously about the prospect of atomic bombs, believed they were too far in the future to be relevant to the war effort (Farmelo). This belief came largely from his science advisor, Professor Friedrich Lindemann (later Lord Cherwell), a man who seldom let scientific data get in the way of his opinions. However, by mid-1940 several UK labs had made important findings in regard to uranium. For example, in March 1940 Otto Frisch and Rudolph Peierls at the University of Birmingham saw how to turn a mass of U235 into a bomb: by slamming about half of the required 'critical mass' down a gun barrel into the other half. This would rapidly put enough U235 nuclei close enough together for their natural radioactivity to create an explosive chain reaction. And the required mass, Frisch and Peierls determined, was a matter of pounds, not tons as had earlier been thought (Farmelo).

They sent a memo about their findings, and their concern about potential Nazi progress, to a senior government scientist, Henry Tizard. By mid-April 1940, a British government uranium committee code-named 'MAUD' was beginning to coordinate urgent uranium research

---

1. See 'What Really Happened' (Dec 1940).

at labs throughout the UK. This independent verification of findings indicated that researchers in both the US and UK were on the right path. But it also suggested that their counterparts in Germany might be on the same path. In May, *The Times* of London, drawing on information from *The New York Times*, reported that German scientists had been ordered to drop everything else to work on atomic fission (Farmelo).

During the summer of 1940, Henry Tizard led a mission to the (still officially neutral) US in which they shared data and technology secrets which the British had developed (Farmelo). In addition to information on radar and bomb fuzing, the Tizard mission also shared the latest uranium findings, and the British concern about a Nazi atomic bomb. The British had learned that the Germans were collecting all the heavy water they could get from the Vemork chemical plant in Norway. This facility was the world's only industrial source of this very rare substance. Heavy water contains a heavy isotope of hydrogen and its sole use was in nuclear physics research. The British believed it could be used to help control neutrons in a nuclear chain reaction. So the Germans' interest in heavy water suggested they believed that too (Farmelo).

This British technical and intelligence data further spurred Mendenhall to advocate a major development effort. Mendenhall and his committee compiled data from various US physics researchers, and combined that with additional and corroborative data from their British counterparts. By October 1940 he was able to craft a compelling story for decision-makers, explained in 'sentences short enough even a politician can understand' as one of Mendenhall's staffers reportedly quipped.

Mendenhall reported that neutrons from radioactive decay can cause uranium 235 nuclei to split, releasing energy and more neutrons. A mass of U235 of 100 to 200lbs would likely undergo an enormously explosive chain reaction as soon as it was put together. Such a mass could be assembled by some sort of gun firing a uranium bullet into a uranium target. U235 was scarce and difficult to separate from the predominant U238. There were several techniques that could accomplish this, Mendenhall reported, although every method would require huge, costly facilities and none would yield quick results.

Mendenhall further reported that the predominant isotope, U238, was also promising. It was much less prone to fission, but there were indications that bombarding it with neutrons would create a new element, plutonium, which was likely to prove fissile. If that were the case, then useful quantities of plutonium could perhaps be obtained using more or less conventional chemical separation techniques to extract the plutonium from the uranium. The challenge would be to build a facility to irradiate large quantities of uranium to create plutonium. Fermi and Szilard were developing an 'atomic pile' to demonstrate that a neutron-emitting chain reaction in natural uranium could be sustained and controlled. Such a reactor would turn some of the predominant U238 into plutonium. German scientists may know all of this, Mendenhall's committee stated, and the Nazis may already be acting on it.

A Chair Not Taken

As noted, Lyman Briggs was the White House's initial choice to chair the Uranium Committee, but his health issues in late 1939 led to the appointment of Walter Mendenhall instead. In mid-1940, Briggs had recovered and joined the Committee as Mendenhall's deputy. Lifelong bureaucrat that he was, he 'approached the unknown territory of a nuclear fission weapon with caution. His deliberate pace angered some of the leading scientists involved, including E.O. Lawrence, I.I. Rabi, and Leo Szilard' (Landa and Nimmo). Szilard and his colleague Eugene Wigner believed that the 'swimming in syrup' bureaucratic pace at which Briggs tried to proceed posed serious delays for the project (Szanton). Fortunately, Mendenhall quickly became aware of Briggs' attempts to keep things at a normal, deliberate, peacetime pace, and worked directly with the various scientists who were themselves hustling to bypass the Briggs Barrier. One duty Mendenhall left to his deputy Briggs was liaison with the British MAUD Committee. In that capacity, Briggs received minutes from MAUD Committee meetings discussing findings about uranium properties and bomb possibilities. Later in 1940, British physicist Mark Oliphant asked Ernest Lawrence at Berkeley what he thought of 'the cyclotron data in the latest MAUD report'. But Lawrence had not seen the report. This was because Briggs, the cautious bureaucrat, had locked the sole copy of the MAUD report in his safe, sharing it with no one (Rhodes 1986). Oliphant reportedly called Briggs an 'inarticulate and unimpressive man' (Rhodes 1986). Oliphant soon persuaded Mendenhall to immediately ensure that the UK's MAUD information be promptly shared with the Uranium Committee members. After the war, Oliphant observed, 'Thank goodness he was not in charge of your Uranium Committee. Who knows how much his bureaucratic foot-dragging would have delayed the bomb.' (See 'What Really Happened' [1940].)

Review, Report, Repeat

Mendenhall's Uranium Committee report of October 1940 came just a year after the committee's inception.[2] By this time, Vannevar Bush was head of the National Defense Research Committee. FDR created this to coordinate scientific efforts supporting the coming military confrontation (Rhodes 1986). A cautious scientist, Bush paid close attention to Mendenhall's Uranium Committee information, and recognised its potential importance. But he was not wholly convinced. So he did what government officials do when they are faced with making a decision without all of the 'comfort data' they'd like: he set up a 'Blue Ribbon' committee. He asked the prestigious National Academy of Sciences (NAS) to review the Uranium Committee's report (Rhodes 1986). This 'independent panel of peers' headed by Arthur Compton reviewed the Uranium Committee's work and reported that atomic bombs were indeed feasible. Also as the Uranium Committee had estimated, Compton's group said it would take about four years to construct the processing facilities and obtain sufficient bomb fuel (Rhodes 1986).

In December 1940, the same month Bush received this NAS report, Fermi and Szilard succeeded in demonstrating a self-sustained nuclear chain reaction. Because they were using natural uranium, which contains very little of the fissionable U235 isotope (less than 1 per cent) they had to use a large mass of uranium – over 5 tons. The team was reasonably sure they could control the chain reaction by adjusting the positions of control rods made of cadmium, which harmlessly absorbs neutrons. These rods were necessary because when a U235 nucleus absorbs a neutron and then splits apart, it releases more than one neutron. If there had been no way to 'filter out' some neutrons, the chain reaction would generate more and more neutrons, and fission events, leading to an uncontrolled build-up of heat. Fermi and Szlard's calculations proved correct; the reaction never got out of control. This was fortunate, because the experiment took place in the basement of Schermerhorn Hall at Columbia University in Manhattan. (Therefore, the Columbia administration may not have been unhappy when Fermi moved to the University of Chicago the next year.)[3]

The ability to initiate a chain reaction in natural uranium also demonstrated that it would be feasible to 'cook' U238 into plutonium, the alternate bomb fuel. Indeed, it was in the same month, December 1940, that Glenn Seaborg and colleagues at Berkeley determined that plutonium

---

2. See 'What Really Happened' (Nov 1941).
3. See 'What Really Happened' (2 Dec 1942).

239 created from neutron bombardment of U238 was indeed fissile. This meant that by 'breeding' plutonium from the relatively abundant U238, there could be potentially far more bomb material available than just the scarce U235.

In response to these exciting new developments at the end of 1940 Vannevar Bush promptly – asked for another NAS study. This one was to look at the engineering aspects of uranium separation and plutonium extraction (Rhodes 1986). He received this report in February 1941. This study reaffirmed the feasibility of atomic bombs based either on uranium or plutonium, and now predicted that they would take about three years to produce.

In April 1941 after a third review study, this one including ordnance experts such as Harvard chemistry professor and explosives specialist George Kistiakowsky (Rhodes 1986), Bush was finally convinced. He and Mendenhall met with FDR. They told him that not only 'the eggheads' but engineers and ordnance experts were convinced that the US could build almost unbelievably powerful atomic bombs in three or four years. Based on their knowledge of their German counterparts, the US and UK physicists also believed that the Nazis could already be developing these weapons.

In April 1941 the US was not yet at war, but was clearly preparing for it. In his 'Arsenal of Democracy' speech the previous December, FDR had made the case that it was in America's interest to aid Great Britain and make sure the Axis did not prevail. So far this was (officially) limited to materiel assistance, and America's factories and shipyards were busy producing war supplies for use against the Axis. But even while it professed to be neutral, America's own 'peacetime' military forces were growing from half a million to nearly two million (Census Bureau).

Chapter 3

The Manhattan Engineer District

Although not yet at war, by mid-1941 FDR was making major defence investments, and a programme to create a new form of munition was fully in line with his other actions. He asked Vannevar Bush to recommend a plan for managing that programme (Rhodes 1986). Bush suggested that the entire effort, involving top-notch scientists and leading academic labs, as well as the construction and operation of enormous industrial facilities, be run by the Army's Corps of Engineers. Since Revolutionary times, Army Engineers have had dual military and civilian roles, supporting troops on the battlefield, and building dams, flood control and navigation improvements for communities nationwide. Tapping the Corps for this new programme was therefore a logical choice, and on 2 June 1941 Roosevelt concurred with Bush's suggestion.

The army's Chief of Engineers selected Colonel James Marshall as Project Director in June 1941 (Rhodes 1986).[1] Marshall was District Engineer in Syracuse NY at the time and was responsible for building new army bases and airfields in New York and Pennsylvania to help accommodate the swelling army. He set up his organisation in Manhattan, where it was easier to find office space than in Washington. As further cover, he styled himself leader of the so-called Manhattan Engineer District (Rhodes 1986). There were already dozens of geographically-defined Engineer Districts around the country, so the establishment of this one more was not ostensibly noteworthy. While not a major base for troops, New York City had about half a dozen Army forts within and on the outskirts of the city. These were mostly coastal fortifications (some dating to Dutch colonial times) guarding New York Harbor and the approaches to it. So it made sense there would

---

1. See 'What Really Happened' (Jun 1942).

be an Engineer District apparently dealing with military structures in New York.

A civil engineer by training, Colonel Marshall was sceptical of the scientists' assertions about the feasibility, the cost and the time requirements of the several isotope-separation technologies they were proposing (Atomic Heritage). So he proceeded at the deliberate pace the Corps of Engineers has used for generations in building dams and other pet projects in the districts of influential congressmen. Marshall identified a suitable site near Oak Ridge, Tennessee for some of the uranium separation plants, but then held off acquiring the land until the design of the facilities was more advanced (Atomic Heritage). He also declined to make any serious preparation to construct the isotope-separation facilities until the physicists could tell him which of the several processes (thermal diffusion, centrifuge, gas diffusion, electromagnetic separation) would work best.

A sceptic turned enthusiast, Vannevar Bush became increasingly frustrated with Colonel Marshall's slow pace. He was reported to have remarked to Mendenhall, 'you can always tell the Corps, but you can't tell them much'. At the end of 1941 US entry into the war seemed ever more likely. Bush and Mendenhall were convinced that America had to be the first to acquire this new weapon, and as soon as possible.

In mid-November 1941, Bush met with War Secretary Stimson to 'suggest' that the 'Manhattan Project' as the bomb effort was then being called, needed a kick in the pants from a new leader. Stimson tasked his staff to identify suitable 'ass-kicker' candidates. Colonel Leslie Groves quickly came to the top of the list. Some said this was because his bellowing could be heard at the War Office building right across the Potomac from where Groves was supervising a big construction project. He was in charge of putting up a new building of unprecedented size, and decidedly odd shape, on the Virginia side of the river; it would house the 40,000 personnel of the expanding War Office. He was working from design drawings that had been developed in days and weeks rather than months and years, and his team had to accommodate major on-the-fly design changes, such as how many storeys it was to have (Vogel). Even before taking on the construction of what is now called the Pentagon, Groves had a reputation as 'a doer, a driver, a stickler' (Fine and Remington). Five months into the Pentagon job, it was already clear that Groves was not a man who worked at a deliberate pace, and was a man who could handle large, complex undertakings.

Over the first weekend in December, Stimson was still weighing the virtues of diverting Groves from one important job to an even more important one. By Monday, 8 December, with the help of the Japanese and their attack on Pearl Harbor, he'd made his decision. The Manhattan Project needed to move as fast as possible, and Leslie Groves was the man to make that happen. 'Colonel Marshall is a highly competent officer, but with the new circumstance after 7 December, the bomb project needs to get out of first gear, and all those involved in the project need to see how serious we are about this,' Stimson told a deputy.

On 11 December 1941, with two wars on his hands, FDR concurred in the appointment of now Brigadier General Leslie Groves to invigorate the Manhattan Project.[2] Groves did not disappoint. Within days of his appointment, he excused himself from a meeting with several of his superiors so that he could catch a train to inspect the Oak Ridge site that his predecessor had kept on hold. Groves ordered it purchased a few days later (Rhodes 1986). In the same timeframe, he acquired 1,250 tons of exceptionally high grade uranium ore from the Shinkolobwe mine in the Belgian Congo, which the mining company had shipped to New York the previous year to keep it out of Axis hands. A large portion of the planet's above-ground uranium sat in a warehouse in Staten Island, NY about 10 miles from Times Square; no one in the US government had expressed interest in buying it. Groves also promptly contracted with the mining company to buy essentially all of the uranium they could mine and ship across the South Atlantic to the US (Zoellner). The Shinkolobwe mine yielded the richest ore (80 per cent uranium) of any place on the planet (Marnham); without the Staten Island stash and subsequent shipments from the Congo, the Oak Ridge isotope separation plants would have been starved for feedstock (Nichols).

Additional stocks of uranium came from mines in northern Canada, and from West Virginia warehouses: since at least the 1930s, several popular lines of chinaware (Fiestaware among others) had used significant proportions of uranium oxide in their brightly-coloured glazes. This ended in the early 1940s, not out of any concern about the health effects of eating off radioactive plates, but because the Manhattan Project requisitioned all of the manufacturers' stocks of uranium (ORAU).

There were increasing indications from the scientific community that the Germans were actively pursuing the bomb and might be ahead of the Allies. As Vannevar Bush's deputy James Conant observed, 'three months' delay might be fatal' (Rhodes 1986). There was no more time to waste evaluating whether centrifuges, thermal diffusion, gas

---

2. See 'What Really Happened' (Sep 1942).

diffusion or electromagnets would work best to separate the uranium isotopes. Money was not nearly the concern that time was. At the urging of Mendenhall and Bush, Groves began right away to pursue several separation methods in parallel, and also to pursue the plutonium pathway to a bomb. This involved designing and building industrial scale 'uranium piles' wherein U238 would be irradiated to form plutonium. This would then be extracted in yet other specialised facilities the likes of which had never before existed. Within a remarkably short time after Leslie Groves took over, several huge uranium 'piles' (what are now called nuclear reactors) were taking shape to serve as plutonium production plants. Also being built were giant facilities for gaseous diffusion, and electromagnetic separation of U235 from U238.

The once-promising gas centrifuge method was scientifically sound, but despite considerable effort, there were major problems in engineering reliable industrial-scale machines. This method therefore was not used for large-scale uranium separation – during the war. However, shortly after the war several design innovations were made, enabling the construction of highly effective large-scale centrifuge separation facilities (Kemp). It has been speculated that these design innovations could well have been developed earlier – it was a matter of insight, not an absence of technological capability. Had this much more efficient uranium separation technique been available in the 1940s instead of the 1950s, perhaps sufficient enriched uranium for the bomb could have been available much sooner than 1945 (Benford). The difference this might have made is fuel for interesting speculation.

Groves recognised that it was not important which uranium separation technology or which bomb fuel was best, most efficient or least costly. What was by far most important was having the bomb before Hitler did. For the next three years, driven by this perceived existential threat, Groves' Manhattan Engineer District accomplished complex design, construction and development tasks on a scale no one had considered possible before the war started. From the secret Oak Ridge facilities hidden in the forested mountains of Tennessee, quantities of 'enriched' uranium (containing far more U235 vs U238 than in natural ore) were becoming available. From the nuclear reactors on the sprawling Hanford Engineer Works in eastern Washington State's cold desert, quantities of plutonium created from U238 were also becoming available.

And at Los Alamos, a secluded mountaintop laboratory complex in northern New Mexico, physicist J. Robert Oppenheimer led hundreds of the smartest scientists and engineers in the US in figuring out how to package these materials into bombs the likes of which no one had

> ### GOD AND THE GENERAL
> General Groves was always scrupulously careful in managing every project dollar, but there were a great many of them. As the colossal isotope factories at Oak Ridge, and the plutonium-breeding reactors at Hanford took shape, a staffer claimed after the war that he had seen a sign somewhere in the project offices that read:
> *Building Things The Way God Would, If He Had The Money.*

ever seen. At its peak, the Manhattan project employed about 130,000 people, very few of whom knew the objective of the overall project.

A Nazi Nuke?

Leslie Groves, however, never lost sight of the objective of making sure the US was the first to possess an atom bomb. Information about German progress on atomic weapons was distressingly scarce. As Vannevar Bush told Roosevelt early in the war: 'We still do not know where we stand but it is possible that Germany is ahead of us and may well be able to produce superbombs sooner than we can' (Takaki).

There had been worrying reports: in occupied Norway, the Germans had fortified the Vemork heavy water plant. It appeared that the Germans sought to amass enough of this material to use as a neutron moderator in breeding plutonium from uranium (Dahl). (The pile Fermi and Szilard built at Columbia University used highly purified carbon [graphite] for this function.) In September 1941 Werner Heisenberg, Germany's principal nuclear physicist, had met with his mentor Danish physicist Niels Bohr before the latter escaped to Britain. The particulars of this meeting are still to the present day a matter of, well, uncertainty. Heisenberg may have boasted that Germany was well advanced in its effort to develop atomic weapons, and apparently disclosed to Bohr that he had designed a reactor which would use heavy water to create plutonium from uranium (Rhodes; Cassidy). Once in the US in late 1943, Bohr is reported to have met with Oppenheimer and told him that Heisenberg was working on an atomic bomb. Bohr showed Oppenheimer a sketch of the reactor that Heisenberg had described in his meeting with Bohr (Bethe). Had Heisenberg become Hitler's Oppenheimer, leading the German equivalent of the Manhattan Project? The US could not be sure. (As physicist Arthur Compton wrote after the war, the US was so unsure that some of the troops landing in Normandy on D-Day carried Geiger counters to detect radioactive debris [Takaki].)

By mid-1944, as the Allies prepared their landings in northern and southern France, Groves acquired the added responsibility for Project Alsos. This effort would deploy technically knowledgeable teams following close behind the advancing Allied armies; these experts would seek out scientists and facilities associated with the German atomic bomb programme.

In mid-June 1944, Alsos personnel entered Rome right behind US troops. Over the next few weeks they interviewed Italian scientists about their knowledge of German nuclear research. This information suggested that the Germans had not advanced far at all. But this was far from a definitive assessment (Jones).

Then in September 1944, Alsos personnel in liberated Belgium discovered that 1,200 tons of Belgian Congo uranium had been shipped to Germany earlier in the war (Groves). This was a concern. In November, Alsos personnel discovered a German nuclear research facility in Strasbourg (a French city the Nazis had annexed early in the war). However, the documents found in Strasbourg strongly indicated that the Germans had not developed any effective technique for separating uranium isotopes. Based on this, the Alsos team reported that Germany therefore could not possess uranium weapons, nor even be close to having them (Goudsmit).

Unless the Germans had discovered some wholly new processes, any facilities for nuclear fuel production would have to be very large and would have enormous resource requirements. The uranium separation plants at Oak Ridge consumed about 1 per cent of all the electrical power in the US (C. Reed). The series of plutonium-breeding nuclear reactors near Hanford used a good portion of the Columbia River for cooling water and hydropower for plutonium separation. If the Germans had built any facilities analogous to those of the Manhattan Project, they should have been obvious to the numerous and widespread aerial reconnaissance missions flown over the Reich. Unless they were underground. But putting such huge facilities underground would have represented an even more monumental effort by German industry, which the team considered unlikely. (The Alsos team was unaware at that time of the miles of caverns the Nazis had excavated near Nordhausen and were then using for production of the V-2 and other advanced weapons.) Overall, the Alsos team did not believe Hitler had or would soon have either a uranium or a plutonium bomb. Still, the credibility of this and any Allied intelligence assessments was shaken a few weeks later when Hitler launched a major offensive in the Ardennes that caught the Allies almost totally by surprise (Cole).

As the Allied armies countered the German surprise in the Battle of the Bulge in December 1944 and January 1945, the Alsos team reviewed what they knew or thought they knew. Perhaps the Strasbourg documents describing only a rudimentary nuclear programme were disinformation, 'plants' intended to be found. Still, if the Germans did possess atomic bombs, this Ardennes offensive would have been the time and place to use them. So maybe the Germans did not have them yet, but would soon. Or, maybe the Nazis were preparing to use atom bombs against cities or ports. Was it possible that they had built a V-2 type rocket large enough to deliver an atom bomb, say to Brussels? What had the Nazis done with the thousand tons of Belgian Congo uranium they had seized early in the war along with all the output of the Joachimstall uranium mine in Czechoslovakia? Had the Germans developed – and concealed – the means to separate U235 on an industrial scale, or to breed and extract plutonium? The Alsos team still did not think it was likely that the Germans had the bomb, but based on the limited data they had in late 1944, they could not rule it out. There were indeed persistent rumours of a German bomb (Hinsley et al.).

Chapter 4

How to Use the New Weapon

When Leslie Groves took over the Manhattan Project days after Pearl Harbor, the prediction from the NAS studies, and similar analyses by the British, was that it would take from two and a half to four years to produce deliverable atomic weapons (Rhodes 1986; Farmelo). Two and a half years later, Groves told Army Chief of Staff George Marshall that the first weapons would be available right about at the three-year mark (Frank). That would be roughly 1 January 1945.

With that projection in hand, in September 1944 a Top Policy Group grappled with plans for using the bombs. War Secretary Henry Stimson, Navy Secretary James Forrestal, Army Chief George Marshall, Navy Chief Ernest King and Secretary of State Edward Stettinius discussed various options. Despite the assessment by the Alsos science intelligence team, they could not rule out that Germany might yet acquire the bomb before the end of the war. Few planners doubted that if Hitler acquired a nuclear weapon, he would promptly use it. All agreed that if Hitler used an atomic bomb, retaliation in kind should be as immediate as possible.

As to Japan, the belief was that they too might have a nuclear weapons programme. Japan was known to have some European-trained nuclear physicists, such as Dr Yoshio Nishina, who had acquired cyclotrons from UC Berkeley before the war (Raghheb). Professor Hikosaka Tadayoshi had published an 'atomic physics theory' in 1934 in which he speculated that the huge energy contained by nuclei could be used to build weapons (Wilcox). Yet while there was some concern about a Japanese A-bomb effort, Allied experts did not believe they could be very far along in that technology. Nor did they believe the Japanese had access to the quantities of uranium they would need (Low; Adelstein). After the war, this intelligence was found to be incorrect; in the early 1940s, Japan had found uranium deposits in Korea, a land they had annexed decades earlier (Berger). Nonetheless, it was true that the Japanese atom bomb programme had not progressed very far. In part this was because Japanese scientists had not believed that they or any other nation could develop a bomb in less than about 10 years (Rhodes 1986).

So aside from the unlikely scenario of the Axis using an atomic weapon first, prompting an in-kind response as swiftly as possible, the senior US officials considered how to use their atom bombs when they became available. Should they drop one on an Axis target without warning, or should there be some advance notice, or even a demonstration of some kind? Several of the atomic scientists, including Glenn Seaborg and Leo Szilard, strongly advocated a non-lethal demonstration of the bomb, hoping to induce enemy surrender without having to use the new weapon (Rhodes 1986).

The planning group considered the idea of a demonstration in the North African desert or an uninhabited Pacific island, perhaps in some way under the auspices of a neutral country such as Portugal, Sweden or Switzerland. But virtually no one in Washington believed that Axis observers could even be persuaded to witness such a demonstration, let alone that it would induce their governments to capitulate. Anyway, as advisor Arthur Compton recalled, 'We could not afford the chance that a demonstration bomb would be a dud' (Takaki).

Another argument for demonstrating the weapon before making use of it in combat was based on the experience in regard to chemical weapons. The thinking was that both Axis enemies had chemical weapons, poison gasses such as chlorine and mustard and phosgene, like the Germans used in the First World War. But they had held back on using them in this war. (Though the Japanese used gas against the Chinese in the 1930s [Tanaka].) Given the 'restraint' shown by the Axis, the argument ran that for the sake of world opinion the Allies should exercise some analogous restraint – at least insofar as staging a demonstration of this new weapon before using it.

But the counterargument was that the 'restraint' by the Axis had nothing to do with humane considerations. The Germans and Japanese, the reasoning went, were surely aware that the Allies also had those poison gasses available. Therefore, Berlin and Tokyo probably believed the Allies' ability to retaliate with equally deadly gasses outweighed any short-term battlefield advantages the Axis would gain by using them. The thinking in Washington was that the enemies' non-use of chemical weapons was primarily a matter of the deterrence of mutual lethality. The argument that the bombs should not be used out of humane considerations like those of the Axis therefore did not bear scrutiny. As the text box shows, the situation was more complicated.

> **SELF-DETERRENCE**
>
> We now know that in the late 1930s, German scientists discovered an entirely new class of poisons. Unlike the choking and blistering agents such as chlorine and mustard gas of the previous war, these new poisons killed very rapidly by interfering with the nervous system. It was said that the first symptom of exposure to these 'nerve agents' was 'when the guy next to you dies'. During the war, the Nazis prepared artillery shells filled with these new poisons, 'nerve gasses'; but they never used them. Reportedly, the Nazis took the absence of information about these chemicals in the international scientific literature as a sure sign that the Allies had also discovered them and were suppressing discussion of them to keep them out of German hands. Thereby convinced that the Allies could also retaliate in kind with these potentially devastating nerve gasses, the Nazis deterred themselves from using them. But the real reason these chemicals were not discussed in the Allies' scientific literature is simply because they were unknown. Neither the US nor UK knew about these nerve agents until after the war (Borkin; Everts). Thus, throughout the war, Hitler had a superweapon that he did not use. Perhaps some so-called 'alternative historian' will someday write about the mass nerve gas attacks on London.

As to the idea of an advance warning of the use of the atom bomb, the Top Policy Group noted that no warnings were being issued in advance of the mass conventional and incendiary bombings of German and Japanese cities that were by then becoming routine. Such horrific attacks were killing civilians by the tens of thousands. Still, the concern was that this new device would be so clearly *sui generis* that the Allies needed to give the Germans and the Japanese the opportunity to avoid experiencing it, lest the post-war world view America as even more barbaric than these enemies. To be sure, there was a risk involved with advance warning: the enemy might move Allied POWs and/or concentration camp inmates into cities they thought were likely to be atomic targets. But most enemy cities were already conventional or incendiary targets, and prisoners had apparently not been moved en masse into cities. With an eye on how post-war histories would be written, the war planners reached the compromise agreement that some sort of non-specific warning or ultimatum about 'impending utter destruction' should nevertheless be issued just before an atomic strike.

Would such a warning prompt an Axis surrender? As of late 1944 there was no doubt among the Allies that both Germany and Japan were being defeated. Rational leaders in both those nations would have already called it quits and saved their nations further bloodshed and destruction. But rationality was scarce in both Berlin and Tokyo. All indications were that in both countries, both government and military were prepared to fight to the bitter end (Weinberg). It was unlikely that any mere ultimatum from the Allies would alter Axis thinking.

If an ultimatum produced no results, would the US actually use the new weapon? As noted, the Manhattan Project was stimulated by concern that German scientists were developing an atomic bomb. There was not universal agreement, but it was a widely held assumption from the beginning of the bomb project that the US would indeed use atomic weapons against Germany (Weinberg). As Oppenheimer put it after the war, 'Atomic bombs were meant to be dropped, as soon as they were ready, on whatever enemy targets remained' (Gaddis 2005). After the war, General Groves wrote that FDR told him in 1944, 'If the European war was not over before we had our first bombs, he wanted us to be ready to drop them on Germany' (Takaki).

There was also the Soviet Union to think about. The people and the land of the Soviet Union had suffered immensely from the Germans, and the Red Army had incurred huge losses in battling their way westward from deep inside Soviet territory to their current lines in Eastern Europe. Stalin had pressed for a second front since well before the US and British Commonwealth forces landed in France. He suspected that the English-speaking nations were intentionally letting the Germans shed communist blood. What would it do to post-war relations if it turned out that the US had possessed a war-ending weapon, but did not use it while Red Army troops continued to die fighting the Germans? For that matter, how would the American public react if their government held off using a war-ending weapon while American soldiers, sailors and airmen continued to die in Europe and in the Pacific?

What's more, did Stalin know about the bomb? The FBI and Leslie Groves had worked hard to maintain tight security, but there were many known past, and suspected, communists in and associated with the US and UK scientific community. The extent of Soviet espionage penetration was uncertain, but the US could not assume that Stalin was unaware of the bomb project (Rafalko). (After the war, the US found out that there were indeed several highly productive Soviet agents inside Los Alamos and elsewhere in the Manhattan Project [Dobbs].) So the Allies could not defer using the bomb and then hope to persuade the Soviets that it hadn't been ready in time.

There was also the long game to consider. After the Axis was gone, the USSR would remain. An ally of necessity for the last few years, the Soviet Union would likely emerge as a geopolitical rival to the US. As the post-war dynamic took shape, it would be useful for the Kremlin to have a full and concrete appreciation of American capabilities.

The senior war planners then considered whether the bombs would initially be used only against Germany or whether Japan would also experience atomic destruction at the same time. No one knew how strategically effective the atom bomb would be. Would a single bomb be so stunningly destructive as to cause an immediate surrender of both enemies? If the planners believed that a single blast, perhaps hundreds of times bigger than any previous human-caused explosion, would surely bring the war to an end, then clearly the very first bomb should be used as soon as it was available, to end the war and the killing as soon as possible. This first use would presumably be against Germany in accordance with the US-British strategic agreement to defeat Germany first.

Some planners hoped the bomb would be immediately climactic, but no one really thought that was likely (Hastings 2012). Instead, the planners assumed that multiple atomic bombs would be needed. They'd been told that each of the new weapons might release the power of thousands of tons of conventional bombs, but German cities such as Frankfurt, Kiel, Neuss and many others had already experienced bombing raids that each delivered thousands of tons of bombs (Davis). These cities had sustained the damage, and the casualties, and had kept functioning. In July 1943 for example, more than 700 RAF bombers dropped explosive and incendiary bombs on Hamburg, causing a firestorm with 150mph winds and 1,500° heat that killed over 40,000 Germans (Frankland and Webster). Despite the horrendous loss of life and the extensive damage, Hamburg continued producing armaments for the war machine, and it was the object of dozens more bombing raids in 1943 and 1944.

The Allied planners recognised that the enemy might see the detonation of an atom bomb as not all that different from the conventional and incendiary bombing they were already experiencing. To the military mind, using a single bomb dropped by a single plane was in many ways more 'efficient' than dropping thousands of them from hundreds of planes. Yet it was not clear that the explosion of an atom bomb would seem to be anything more than just (a lot) bigger than that of a conventional bomb. The Manhattan Project scientists knew that it would be physically quite different, with intense nuclear radiation never before experienced on Earth, and temperatures higher than any chemical explosive can produce. But would that matter? 'Rubble's rubble; dead's dead', one military planner was reported to have said. Another noted, 'Humanity's capacity for violence is exceeded only by their capacity to endure it'.

Of course the objective was not to make the Germans or the Japanese endure but to prompt them to end the death and destruction. To that end, the planners wanted each detonation, if more than one were needed, to have the maximum psychological effect on the enemy nations. That effect would be enhanced, the planners believed, by using the bombs in relatively rapid succession.

As Rear Admiral William Purnell, an advisor to General Groves, saw the scenario, it would very likely take at least two bombs to make the enemy surrender: the first would show what atom bombs could do; the second, after a few days' pause to give the enemy time to assess the situation and decide accordingly, would show that there were plenty of them (Norris). Under this scenario, it was important that neither enemy sense how limited the supply really was, lest they be tempted to try to somehow 'tough it out'.

Some planners gave strict priority to the Allied strategy of defeating Germany first. Their approach would have been to make Germany the sole target of however many atomic bombs it took to obtain surrender (Gaddis 2005). Bombing of Japan would then follow, if needed. After some debate, however, the Policy Group reasoned that dropping bombs on Germany and on Japan as nearly simultaneously as possible would reinforce their strategic effects in both theatres. 'Convince the Krauts and the Japs we have plenty of these things', was the sentiment. 'Plenty' was then codified as eight. That is, the atomic bombing campaign would begin as soon as there were eight atom bombs available or nearly so. On the first day of atomic bombing, one bomb would be dropped on Germany and one on Japan. There were to be three more pairs ready to fall at intervals of three to five days (weather permitting).

The Policy Group's rationale paralleled that used by the Germans in regard to their V-2 terror weapon. No V-2s were used until the Nazis had assembled a considerable stockpile of them so that they could be used in rapid succession, seemingly drawing on an endless supply (J. Neufeld).

Under the terms of their 1943 Quebec Agreement on nuclear cooperation, Roosevelt and Churchill had agreed that they would not use nuclear weapons without each other's consent (Department of State 1970). Accordingly, the Americans' Top Policy Committee consulted with the Combined Policy Committee's representatives from the UK and Canada, which had also assisted with nuclear research and with supplies of uranium and polonium (used to help initiate nuclear explosions).

Having endured German bombing of their cities during the Blitz, and later the indiscriminate reign of destruction from the V-2s, the British had no reluctance to see the new weapons used on Germany, as well as on Japan. Thus the Combined Policy Committee concurred in the approach to target German and Japanese assets simultaneously.

To determine specifically where to use the bombs, a targeting committee identified several criteria. Targets should be (1) important assets in a large urban area of more than three miles in diameter, (2) capable of being damaged effectively by a blast, and (3) unlikely to be destroyed by conventional bombing before the atomic bombs were ready (Target Committee). In the case of Japan there were, as of autumn 1944, a large number of targets that had not yet been bombed at all. The B-29 bombing raids then being mounted from Nationalist-held China were limited in number and range; they could reach the southernmost Home Island Kyushu, but not nearly as far as Tokyo on Honshu (Haulman). In Germany however, virtually every potentially valuable target had already been bombed by the RAF, the US Army Air Forces (USAAF) or both. Yet despite sometimes repeated bombing raids, the industries and services in many of these German cities continued to function, and therefore continued to present relevant target opportunities.

The planners added an additional criterion in regard to Germany: the target had to be sufficiently distant from any Allied troop formations so as to minimise risks to them. In part this was a concern about bomber navigation errors, but it also reflected concern that there might be radioactive debris from the blast that could be carried by the wind for some distance from the actual target. There was no experiential data about whether such 'radioactive fallout' would occur and if so, how dangerous it might be at what distance and at what interval. Manhattan Project scientists had looked into the damaging biological effects of radiation, but that work had been focused mainly on health and safety concerns for those working on the bomb project. Few scientists had thought about the idea that a nuclear explosion would disperse radioactive material any great distance from the blast site. Those who had found no precedent, no model to make predictions with. If such fallout did occur, it would not be militarily important in Japan – there would be no friendly troops anywhere nearby. But if an atom bomb blasted a German target, there could be Allied troops not far away. Therefore, the request came down from the planners: how far away should a target be from friendly troops?

From somewhere within the Manhattan Project emerged a SWAG ('Scientific Wild Ass Guess') that radioactive fallout could occur perhaps a few tens of miles away from the blast site. Therefore the recommendation was that the target should be at least 30 miles away from approaching troops. (In an apocryphal version of the story, the addendum was that in the case of approaching Soviet troops, the recommendation was 200ft.) The policy planners in Washington did not necessarily care whether the 30-mile distance was the scientifically-correct safety margin. Since prevailing winds at German latitudes blow toward the east, whatever fallout risk there might be would apply primarily to approaching Soviet troops. Washington would be able to say that they used the best available science to minimise any danger to the valiant godless commie allies. The targeting committee adopted 30 miles as the required safety margin.

Based on all the considerations of target size, importance, and condition, and the 30-mile 'safety' margin, the target committee submitted a list of German and Japanese cities meeting the required criteria. Neither Berlin nor Tokyo was on the initial list; the available records are not clear as to why. There have been several theories.

One contention is that the plan was to spare Tokyo because its destruction would decapitate the Japanese government. Allied planners speculated that instead of causing the surrender of the Japanese military, the loss of the Emperor and his government could have spurred the several far-flung military commands, with their several million troops, to continue the fight even more fanatically (Brooks). Years earlier, Japanese Army commanders in Manchuria had exercised a degree of independence that would have been strongly punished as insubordination or mutiny in Western militaries. Often acting clearly against orders

---

A KODAK MOMENT

Very soon after the first atomic explosion, the US Army found radioactive fallout 200 miles from the New Mexico test site (Ortmeyer and Makhijani). This information did not apparently reach the targeting planners. Sometime later, physicists at Eastman Kodak realised that the New Mexico test had ruined a lot of the company's X-ray film: after many customers suddenly began complaining of fogged films, Kodak found that the ruined films' packaging was contaminated with radioactive cerium. The cerium came from the river water used at a plant in Indiana to process the packaging material for the film. The river water became contaminated when cerium-laden dust fell on Indiana fields and got washed into the river. The radioactive dust began falling in Indiana a few days after the first atom bomb blast in the New Mexico desert, 2,000 miles away (Blitz).

from Tokyo, this Kwantung Army engineered the provocation that led to their complete takeover of Manchuria (Behr). Years later, local Japanese Army commanders violated orders and broke an agreement with the Vichy government of Indochina by seizing a garrison near the Vietnam-China border (Dommen). Given this history of autonomous behaviour by major Japanese commands, Allied planners were said not to be confident that destroying Tokyo would hasten the end of the war. It might even prolong it. Overseas troops might refuse to surrender, and the million-plus troops on the Home Islands might not only refuse to give up, but could also disperse and wage a prolonged guerrilla war against Allied invaders.

Another train of thought connects sparing Tokyo to sparing Berlin. Incinerating Hitler was desirable. Allied planners were beginning to get peace feelers from multiple German sources (Weinberg; Padfield). The Allies believed that there was sufficient dissatisfaction among senior German military and civilian leaders such that the loss of *Der Führer* would allow them to promptly sue for peace. The problem was that there could be no assurance that an atomic bomb would kill Hitler. Because they had taken to launching frequent bombing raids against Berlin, the Allies had to assume that if Hitler were in Berlin, he had available to him a well-fortified underground bunker. They did not know how much time he spent in it. (Post-war documents indicated that by mid-January 1945 Hitler was indeed spending most of his time 30ft under the Reich Chancellery's garden. He lived in a comprehensively equipped, reinforced concrete complex [Read and Fisher] which could have survived an atomic bombing.)

The concern was that if an atom bomb levelled central Berlin, and Hitler climbed godlike out of the rubble a day or two later, the inspirational effect on German morale would be enormous. The scientists estimated that in such a scenario Hitler and his entourage would likely die of radiation poisoning not much later. But the planners realised this might not happen before Hitler's defiance of the bomb had spurred his people to ever more fanatical resistance, potentially prolonging the bloodshed on both sides.

Thus, if Berlin was to be spared because the Allies could not risk making a god of Hitler, then the home of the Japanese Emperor, their living god, had to be spared: the planners wanted to forestall post-war accusations of greater brutality against the non-white enemy.[1] Among the

---

1. See 'What Really Happened' (Early summer 1945).

German targets considered and rejected was Peenemünde, on the Baltic coast. Planners knew this was the site where the first V-1 and V-2 terror weapons had been developed, tested and produced. In 1944 the British mounted a 500-plane raid on the facility, which had the unintended effect of killing almost a thousand Polish civilian workers, and prompting the Nazis to shift V-2 production to a different site (Garlinski). The combination of diminished military value and the risk of high non-German casualties kept Peenemünde off the list.

The targets were listed in order of desired priority, but with the recognition that weather and other factors in the field at the time of the mission could prompt a commander to switch to a secondary target. In regard to Germany, there was a further consideration of Allied politics: if multiple bombs were needed, they should not all be used against targets in eastern Germany, the future Soviet occupation zone. US and UK planners did not want the Soviets accusing them of intentionally wreaking disproportionate damage in the sector they would have to manage.

Chapter 5

Atomic Timing

Besides being massively destructive, the atomic bombs would themselves be quite massive. At five tons, they'd be much heavier and physically larger than the conventional munitions (which were usually one ton or less) carried by Allied bombers. So several of the new B-29 bombers were modified to each carry not racks of small bombs, but a single very special bomb. Other modifications included more powerful engines, and an additional seat for the special weapon expert (Hornfischer). Early in 1944 the US Army Air Force formed the 509th Composite Group to fly the atomic bombing missions against the Axis enemies. The plan was for the Group to train in Utah on the specialised procedures that would be needed to employ the new weapons. When the training was complete, the plan then called for one squadron of about a dozen modified 'Silverplate' B-29s to head to the Pacific for missions against Japan. The other squadron would deploy to Europe to use the new weapon against Germany (Tibbets). In command of this special unit was a highly experienced bomber pilot. Paul Tibbets had flown dozens of combat missions in B-17s into occupied Europe before being assigned to help develop the new, much larger B-29 (Hornfischer). Having known this aircraft almost since its beginning, Tibbets knew intimately what it was capable of. Under his command, the crews of the 509th would develop the skills needed to maximise their chances of successfully delivering their unique (and uniquely massive) payloads and living to tell about it. Early estimates of the bomb's power indicated that the plane delivering it had to be at least eight miles away when it detonated. To accomplish that, Tibbets devised a descending, twisting turn manoeuvre more akin to what a fighter plane would do than a huge bomber. He ensured that all pilots in the group were proficient in the blast evasion manoeuvre, even those who would be flying a spotting or observation role (Hornflischer).

> ## Made In America
> When the Los Alamos scientists were anticipating that the gun-type weapon ('Thin Man') would require a 17ft-long barrel, there was consideration of using the British Avro Lancaster to deliver it. The Lancaster was the only available aircraft that could have carried it without major structural modifications (Atomic Heritage 2014). There were discussions with the Lancaster's designer, and a visit to the Avro production facility in Toronto (Cully). The story goes that when the US Army Air Force Chief Hap Arnold got wind of that, he mandated that an American aircraft be used, even though the most capable US bomber, the B-29, would require significant structural modifications (Groves). After the doors of the forward and rear bomb bays of a test-bed B-29 were replaced with elongated ones spanning both bays, tests with Thin Man dummies showed they would barely fit, and would be challenging to drop reliably (Bowen). However, as the Los Alamos team advanced their understanding of the nuclear physics of the bomb, they were able to shorten the gun barrel to about 6ft. That fitted into the B-29 much more easily.

In accordance with the plan to demonstrate an 'unlimited' supply of atomic bombs, General Groves was directed to coordinate with the 509th to make four atomic bombs available to each squadron over the same 10-day period, and to inform the Pentagon as to the earliest date he could accomplish that.

When would Groves be able to supply eight bombs? As noted above, Groves had two bomb programmes running in parallel. As of September 1944, what was likely to limit the number of bombs were the production rates of the two bomb fuels. The uranium isotope separation facilities in Oak Ridge, Tennessee were beginning to produce usable quantities of U235. This uranium would power a bomb using a conceptually simple gun mechanism: firing one piece of uranium into another to create a single mass of fissile uranium large enough to set off a runaway fission chain reaction. But despite the several hundreds of millions of dollars invested in the Oak Ridge facilities, they were initially able to produce only enough enriched uranium (not 100 per cent U235, but not far short of it) for about one bomb every several months (Frank). One bomb could be available in early January 1945, but another uranium bomb could not be ready until about April or May. Each bomb required about 140lbs of the enriched uranium (Coster-Mullen).

However, the three U238-fuelled nuclear reactors at Hanford were beginning to produce plutonium at a comparatively greater rate (Department of Energy 2002). There would be enough for one operational plutonium bomb by about 1 January 1945. There could be perhaps four more a month later with additional bombs at the rate of about one a week (Frank). It helped that the plutonium bombs would use only a few pounds of plutonium each.

The scientists had no significant doubts that the gun-type uranium weapon would work: firing a uranium 'bullet' onto a uranium target such that one fit neatly into a well-shaped cavity in the other would form a critical mass for a nuclear fission chain reaction. This nuclear process would get going and release enormous energy in millionths of a second, far faster than the thousandths of a second it would otherwise have taken for the uranium to mechanically shatter from the gun impact.

The scientists were not so confident about the plutonium device. Because of different nuclear physics properties, they knew that plutonium would undergo an explosive chain reaction even before a gun device could fully insert a plutonium bullet into a plutonium target. This would blow the device apart prematurely, after only a very limited, and militarily insignificant, release of nuclear energy. That is, it would fizzle. So a more sophisticated 'implosion' mechanism was to be used to initiate a useful nuclear chain reaction in a 'solid' ball of metallic plutonium.

To the human eye, lumps of metal seem to be 'full' of matter, with each atom packed right up against its neighbours. And compared to a ball of foam rubber, a ball of plutonium is much more full of matter. But just as a strong hand squeeze of that rubber ball can make it smaller, and therefore more dense, with its component atoms closer together, a sufficiently strong squeeze of even a ball of plutonium metal can make it smaller by forcing its atoms to get even closer together. In this condition of greater density, a neutron released from a fissioning plutonium nucleus would be much more likely to strike another plutonium nucleus, shattering it, and so on (Sublette 2019).

In essence, compressing a ball of plutonium weighing some dozens of pounds would set off an explosive nuclear chain reaction. Calculations with slide rules and rudimentary room-filling electronic calculators showed that this would only be successful if the initiating compression were very rapid, very forceful, and virtually perfectly spherical. The scientists had some concern that they might not have mastered the challenge of creating the perfectly spherical, simultaneous 'implosion'

> ## Composite CYA
>
> The unit created for the atomic bombing missions was designated 'Composite' because it included not only combat aircraft but also its own transport aircraft (Anonymous 1945). It was also composite in another sense: it included British RAF personnel.
>
> The entire atomic bomb enterprise had featured a high degree of US–UK (and Canadian) cooperation ever since 1940. That's when Britain's Mark Oliphant had persuaded the Uranium Committee Chair Walter Mendenhall to overcome the excessive caution of his deputy Lyman Briggs and ensure that the British MAUD Committee's reports were shared with their US counterparts. As a result, Oppenheimer's team at Los Alamos included a large number of British scientists.
>
> Seconding British fliers to the 509th Composite group was a logical extension of the same Allied collaboration. But the significance of having RAF personnel take part in dropping atomic bombs on Germany and Japan was immediately clear to FDR when Air Force Chief Hap Arnold suggested it. 'After what they've been through our British cousins deserve to have a role in this – and it will cover our backsides; if there is any post-war backlash about this thing, it won't all be directed at us.'
>
> There was consideration of adding to the 'Composite-ness' of the 509th by incorporating British aircraft. The Avro Lancaster was in some ways a better delivery vehicle than the B-29 because its bomb bay didn't need the modifications the B-29 'Silverplates' needed. However, there were performance questions concerning the operational ceiling and speed of the Lancaster. There were also logistical considerations in that adding British aircraft would entail complicating the maintenance and support requirements of the Group. In the end, the RAF liaisons let it be known that involvement of RAF personnel in the atomic operations would be sufficient. As a Pentagon wag is reported to have said, 'Nice move; it lets 'em claim a role, but not too much responsibility.'

needed (Rhodes 1986). So they planned a mid-December test of the plutonium-fuelled implosion weapon in the New Mexico desert.

Groves told Army Chief of Staff George Marshall in October 1944 that if the test was successful, the two squadrons of the 509th Composite Wing could each have three nuclear devices by about 1 February 1945, with one more for each squadron by mid-February. This information helped FDR in his late November planning for a Big Three conference

> **SLIDE RULES RULE**
>
> They were bigger and often heavier than a today's smartphones. They were wooden and had no batteries. They could not have connected to the Internet, if that had existed in those days. Slide rules were 'mechanical' calculating devices, but they could perform a wide array of mathematical operations with precision extending to multiple decimal places. With a slide rule, a scientist of the 1940s (and the 1960s) could figure out powers (e.g. squares and cubes as well as their roots), logarithms (base 10 and natural), and trigonometric functions (e.g. sine, cosine). Hundreds of Los Alamos 'slip sticks' helped win the war.

to be held in early 1945. He had initially been thinking of holding it somewhere in the Mediterranean about two weeks after his fourth inauguration, i.e. in the first week of February (Reynolds). Now, with the imminent availability of weapons that could hasten the end of both wars, he adjusted his thinking and asked the State Department to try to arrange a conference with Churchill and Stalin in early March. As originally envisioned, the conference would discuss planning in regard to anticipated post-war issues. Now it was conceivable that the discussions would themselves be post-war.

An important step toward that goal was the successful test of the plutonium weapon in the New Mexico desert just before dawn on 14 December 1944.[1]

> *And these atomic bombs which science burst upon the world that night were strange even to the men who used them.*
> H.G. Wells, *The World Set Free*, 1914.

General Marshall received the news later that day, and he promptly told President Roosevelt, 'Preliminary estimates are that the blast was akin to the entire payloads of about 2000 B-29s'. Roosevelt was pleased that the huge investment he had authorised in the Manhattan Project was about to pay off. He had kept it pretty well hidden too, although his new Vice President-elect, Harry Truman, had almost blown the lid off it in 1943. Truman had headed a Senate panel investigating huge undisclosed defence expenditures (Nuclearfiles.org). War Secretary Stimson had

---

1. See 'What Really Happened' (16 Jul 1945).

managed to get Truman to back off. Now FDR thought to himself that he'd enjoy telling Harry just what it was he almost blew the cover off of. He'd have to remember to do that one of these days.

Along with news of the successful test, Roosevelt cabled Churchill on 15 December his idea to issue, perhaps at the New Year, an Allied ultimatum to the Axis powers. It would be a joint declaration from the US, UK and USSR demanding the prompt unconditional surrender by the Axis powers and promising their utter destruction if they refused. To Stalin, Roosevelt cabled this same suggestion, along with the news that the US had recently developed a 'new weapon of unusual destructive force'. At the time, the US government was not certain they had wholly kept the Manhattan Project's secret from the Russians, but on the chance that Stalin did not yet know much about it, FDR chose to inform his ally about this important war development while telling him as little as possible.

As these cables coursed among the Allied capitals, unwelcome news came in the next day, 16 December, from the battlefields of Belgium. As noted earlier, Hitler had launched a surprisingly powerful offensive, primarily against US troops in the Ardennes forest. American combat power and expertise, logistical capability and outstanding generalship allowed little doubt about the outcome of the German's desperate offensive. But there was hard and bloody fighting for over a month. Among the few leaders who knew about the atom bomb, the obvious thought occurred to them all: should the bomb be used to win this battle?

No, was everyone's immediate thought, for several reasons: the battle was not on German soil, so the bomb should not be used on the battlefield itself. Also it was doubtful that the Air Force could locate a sufficiently large concentration of German troops or equipment in any small area that would warrant the use of the single huge explosion from the bomb. Moreover, it was clear that the multi-national armies under General Eisenhower were going to win this battle, and it was important for all to see that happen. Finally, the value of the bomb was in ending the war, not in being employed piecemeal in winning specific battles. The strategy agreed on earlier in 1944 was to build a small stockpile of them so that as many as eight could be used in rapid succession if need be. Unless, contrary to Allied intelligence agencies' belief, Hitler had and used an atom bomb of his own, in which case the Allies would rapidly retaliate in kind.

As 1944 ended General Eisenhower's men were fighting hard and dying, but pushing the Germans back out of the 'Bulge'. By 25 January the Allied lines were essentially where they had been in mid-December (Eisenhower 1948) (Map 1, page 40). 'Knowing that a weapon existed that could perhaps save many of those boys' lives was a burden', Ike later wrote, 'but it was balanced by the hope that when sufficient of those weapons were ready, the year 1945 would surely be the last year of the war for all our troops.'

The January Declaration

At that point, it was clear Hitler's offensive had been soundly defeated, and the eastward advance into Germany could begin again. Meanwhile in East Asia, the Japanese were being pushed back in the Philippines, Burma and the western Pacific. It was now time to issue the ultimatum that FDR held in abeyance during the Battle of the Bulge. On 29 January 1945, Roosevelt, Churchill and Stalin simultaneously issued their warning to the Axis. Since the Allies had not themselves yet agreed upon post-war arrangements, the January Declaration offered no details as to how the Axis nations were to be treated after the war. There was only a vague assurance that the Allied occupation of German and Japanese homelands would last only until such time as their peoples had established peaceful responsible governments. But the declaration was firm in its resolve: it demanded that Berlin and Tokyo 'proclaim now the unconditional surrender of all of their military forces. We will accept no compromises. We shall brook no delay. The alternative for Germany and for Japan is prompt and utter destruction.'[2]

As expected, there were no official responses. The Japanese press declared, *'Mokusatsu'* ('Ignore with contempt'). The German press proclaimed *'Dummheit'* ('Nonsense'). After reading the press reports, physicist Kurt Diebner in Berlin contacted Otto Hahn (one of the 'discoverers' of nuclear fission). Both were involved in the German nuclear program that had spurred the Allies' nuclear programme. But unlike the Manhattan Project, Germany's effort was a relatively low priority, much less focused research programme involving at least three separate teams. These were more competitive with one another than collaborative. There is continuing dispute to this day about how it happened that the nation with the head start lost the race (See 'Losing The Race, or Dropping Out?'), but it is agreed that by early 1945, the most advanced of the teams had constructed only a crude uranium pile. This device used heavy water to 'moderate' the neutrons emitted from fissioning uranium atoms but it had not achieved the steady self-sustaining reaction of

---

2. See 'What Really Happened' (26 Jul 1945).

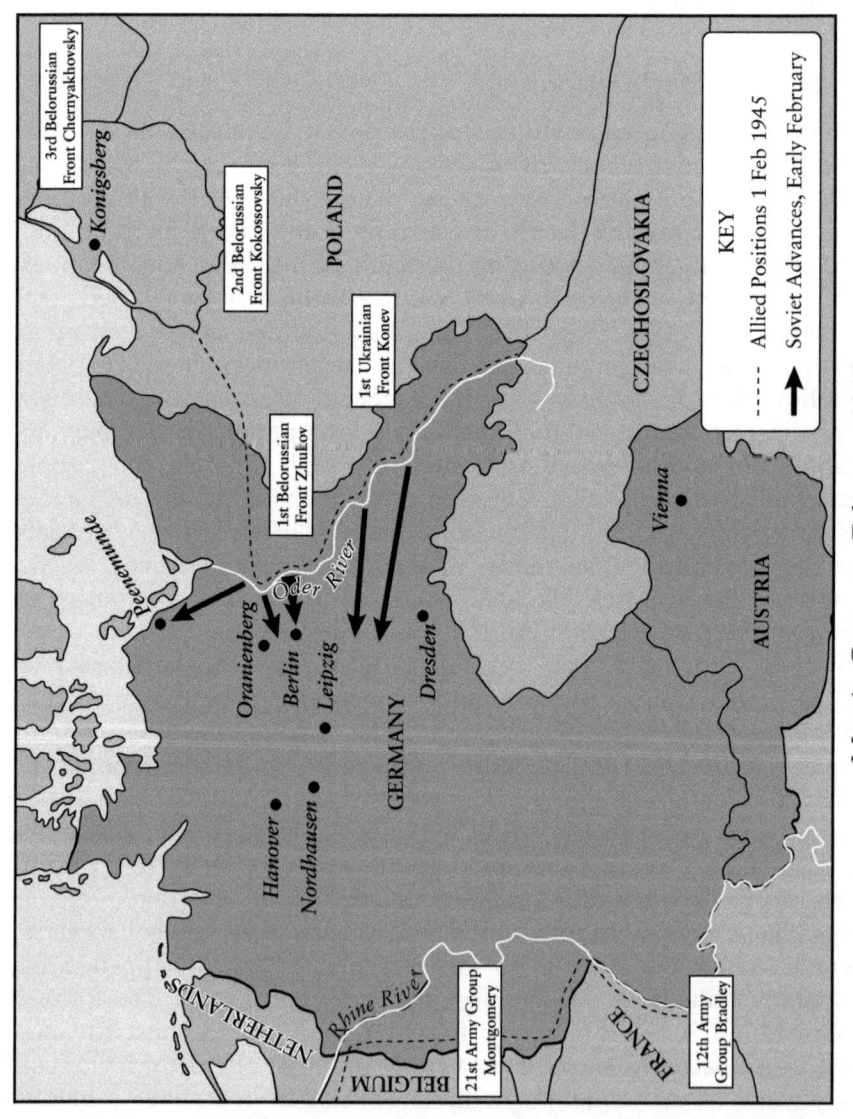

Map 1: Germany, February 1945.

fission–neutron emission–neutron capture–fission as Fermi and Szilard had done years earlier in the US (Koeth and Hiebert). And none of the teams were working on either creating plutonium or enriching uranium, let alone designing a device to use these fuels for bombs. While German scientists such as Hahn and Diebner did know that atomic bombs were possible, they had not believed anyone could have developed them by 1945. Now,

---

LOSING THE RACE, OR DROPPING OUT?

German scientists had discovered fission; the Nazis had seized uranium-rich Czechoslovakia and tons of uranium from Belgium. They were shipping in heavy water from Norway as fast as it could be produced; and leading physicist Niels Bohr told scientists in the US that Germany's Heisenberg had boasted to him about the Reich's nuclear progress. So there was good reason for the Allies to think they were in a race with Germany, a race with the highest stakes.

When the war was over, the Allies were surprised that Germany was nowhere near developing an atom bomb, and they were curious about how that happened. Post-war writings by some of the scientists involved, such as Heisenberg, gave inconsistent accounts of his own and his colleagues' knowledge, and of their objectives (Heisenberg). Manhattan Project scientists asserted that these records showed Heisenberg and others not to have understood basic nuclear weapon principles (Bethe). The declassification after 50 years of some records from 1945 only fuelled further conflicting interpretations (Hoffman; J. Bernstein 1992). Did the German scientists just not 'get it'? Or was the problem lack of support and resources from the government? Or did they intentionally drag their scientific feet to prevent their *Führer* from getting the bomb? A recent analysis suggests that if the uranium which had been allocated among the competing research teams had instead all been available to the one team that did assemble a crude atomic pile, that device would have achieved a sustainable nuclear chain reaction (Conover). Even so, that would certainly still have been a long way from having a bomb.

To be sure, there are claims to the contrary. At least two historians claim to have recently unearthed evidence that the Germans did conduct at least one test of a crude nuclear explosive device late in the war (Karlsch and Walker), but investigation of their remarkable claim has turned up no corroborating evidence (Furlong).

the press accounts of the January ultimatum made them strongly suspect that this was what the Allies meant by 'prompt and utter destruction'.

'We should warn them', Diebner said.

'You want to get word to Göring,' Hahn replied, 'that one day soon one of the thousands of bombers regularly flying over Germany may be carrying an especially big bomb, and it will destroy some German city? Exactly what would you have the Luftwaffe do? They'd probably just have us all shot for not getting the bomb first.'

It is possible, but unrecorded, that analogous conversations occurred among the small nuclear physics community in Japan. Regardless, the US planners had anticipated their enemies' potential thought processes. Neither Germany nor Japan had full control of their own skies, but they did still have air defence capabilities. Accordingly there was no point in making an easy and obvious target of the atomic bomb-laden aircraft. The plane carrying the bomb would be escorted for much of the mission not only by a few instrumentation and observer aircraft of the 509th Composite Group, but also by dozens of conventional bombers from other units. In fact, the special aircraft would be embedded in the larger formation. And the formation containing the atomic aircraft would be just one of several formations heading into enemy territory at the same time. As they neared the atomic target, the conventional bombers would divert toward a secondary target, while the atomic bomb flight proceeded to its intended target. The reasoning was that even if the enemy were expecting an unusual bombing raid somewhere over their territory, there would not be enough time for them to identify and intercept the atomic aircraft once it began its final approach to the target.

Chapter 6

Soviet Planning

While there was some speculation about atomic bombs among scientists in Berlin and Tokyo, the Kremlin was quite sure what was in the offing. Well before the American president had told Stalin about the new weapon of 'unusual destructive power', Soviet spies within the Manhattan Project had informed the Kremlin that a test of the atom bomb was imminent (Rhodes 1986). Like the Americans, the Soviets believed that the advent of the atomic bomb would be a quantum change in the war, but whether the bomb would immediately end it was not clear. Stalin reasoned that he had to act quickly, both in Europe and in Asia.

Europe

By 26 January 1945 Marshal Zhukov's 1st Belorussian Front (with about one million troops) was on the Oder River, 40 miles east of Berlin (Keegan). He had spent the previous two weeks pushing Germans west across Poland, sometimes as much as 25 miles per day (Hastings 2012). Now he informed Stalin that if he were allowed four days to bring up fresh troops, supplies and some new equipment, he could be ready by 1 or 2 February to attack toward Berlin. He told Stalin he could then take the city by mid-February (Read and Fisher). Marshal Konev, commanding the similarly-sized 1st Ukrainian Front (Lunde) on the Oder to the south-east of Berlin (Map 1), said he could be ready to carry the offensive across the Oder in his sector two or three days later (Ziemke). However, Rokossovsky's 2nd Belorussian Front to the north, advancing through Prussia against heavy opposition, had not yet reached the Oder. Military prudence called for having Zhukov and Konev wait at the line of the Oder until Rokossovsky could catch up, so that all three fronts could advance westward without exposing flanks to the German forces. Advancing westward along a broad front would also help ensure that the Red Army took as much German territory as possible, especially given that the Allied European Advisory Commission, set up by the 1943 Teheran Conference, had not yet made its recommendations on delineating zones of occupation (Keegan). However, Stalin specifically wanted the Red Army to take Berlin. He wanted Hitler. He also believed the German atomic bomb programme was based there, and however far they had gotten on the bomb, he wanted that too (Beevor).

Clearly, the end of the war was near. The degree of military prudence that might have been called for in a prolonged campaign was now less important than acting swiftly to be in the best positions when the war ended.

On 29 January, Stalin approved Zhukov's and Konev's plans to resume their push toward Berlin as soon as they had accomplished a few days' of resupply (Glantz 2001). To get to Berlin and Hitler, Stalin was willing to risk exposing his army's northern flank to German forces still in the field to the north-east of Berlin. As if to reinforce the Kremlin's decision, on 3 February, partly in response to the Soviets' request for assistance, 1,000 USAAF B-17s bombed Berlin, especially focusing on rail yards critical to the movement of German troops toward the Russian front (Read and Fisher). In the next few days, as they had planned, Zhukov and Konev resumed their advance to capture Berlin (Glantz 2001).[1]

Japan

Stalin's spies could not tell him how many atomic bombs the US would have, how soon, or where they were to be used. But given that FDR had issued his ultimatum to both Germany and Japan, Stalin recognised that he needed to revise his strategic plan in regard to Japan. The idea of attacking Japan was not new. In a sense, it dated to the Russian defeat in the 1905 war with Japan. But even during the current war, the plan for the Soviet Union to move against Japan dated to very soon after Japan's ally Germany attacked the USSR in June 1941. Japan had not joined in, so the Soviet Union and Japan were not at war at that time. But in December 1941, Stalin told British Foreign Secretary Anthony Eden that he would attack the Japanese about three or four months after he was finished in Europe (Hasegawa). After they defeated the Nazis, the Red Army plan called for spending the next few months

---

1. See 'What Really Happened' (Early Feb 1945).

transporting dozens of divisions (each with about 10,000–13,000 troops) more than 5,000 miles across the Eurasian landmass. To this end in 1943 Stain ordered improvements to the Trans-Siberian Railway (Hasegawa). Once in Siberia, the veteran troops would supplement the units who had spent the war watching the border with Japanese-occupied Manchuria, just in case the Japanese broke their neutrality. The combined Soviet force, plus troops from Soviet ally the People's Republic of Mongolia, would then attack Japanese holdings in Manchuria (Glantz 1983).

At the 1943 Teheran Conference, Stalin affirmed to Roosevelt and Churchill that he would make war on Japan after the Germans were defeated. As if the Soviets were doing the Western Allies a favour rather than acting in their own Eurasian interests, Roosevelt conceded to Stalin's demands for the several Kuril Islands held by the Japanese, the Japanese-held southern half of Sakhalin Island, and access to the ice-free ports of Dairen and Port Arthur on China's Liaodong Peninsula. (Of course, none of these were Roosevelt's to give away.) However, specifics and timing of the Soviet actions against Japan were not settled at Teheran (Department of State 1945). Now at the beginning of 1945 – sooner than he had expected – Stalin needed to develop his battle plan against Japan. His designs on Japanese holdings required that his army get into the fight before the Americans forced the Japanese to quit (Hasegawa).

Although the Soviet-Japanese Neutrality Pact was still in effect, Stalin would have little compunction about abrogating that pact just before he attacked. Neutrality Pact or no, Stalin never wholly trusted the Japanese. Japan's 1904 surprise attack on the Russian fleet at Port Arthur, and their surprise attack at Nomonhan on the western Manchurian/Siberian border in 1939, played at least as much of a role in Stalin's opinion of the Japanese as did their more recent similar actions at Pearl Harbor. Although the Soviets had won the brief 'Nomonhan War', that attempted northward incursion by the Japanese into Siberia had been a major reason Stalin had needed to keep dozens of divisions on guard on that border. After Hitler had launched his own 'surprise' attack on the Soviet Union, Stalin did thin out his forces in the Far East and shift about twenty divisions to the European front to repel the Nazis. But despite intelligence sources assuring him that Japan would next attack south, not north (Andrews), Stalin's distrust of the Japanese was such that he could not risk leaving fewer than about 40 divisions (about 500,000 troops) guarding his 2,600-mile border with Manchuria (Glantz 1983).

In Manchuria (the north-east extension of China) the Soviets' Red Army would face Japan's Kwantung Army. The Soviets had bested them in border disputes at the Battles of Lake Khasan and Nomonhan in 1938 and 1939 (Chen 2009; Teague). And now that million-man Japanese force was much depleted. Many units had gone south to support Japanese operations in south-east China, and into the Pacific theatre against the Americans. Japanese replacement units were understrength, ill-equipped garrison troops, not combat units with proper equipment (Keegan).

Stalin directed his Far East Command to prepare to launch an offensive into the industrial heart of Manchuria in early February. While his military headquarters (the Stavka) was hurriedly planning an early Manchurian campaign, Stalin began diplomatic discussions with one of the two major forces fighting the Japanese occupiers in China. This was Chiang Kai-shek's Guomindang (GMD), the 'Nationalists', the official government of China. (Mao Zedong's People's Liberation Army of the Chinese Communist Party was the other major force.) Stalin invited the Nationalists' Foreign Minister T.V. Soong to Moscow to negotiate an alliance against the Japanese (Holloway).

Even without a formal alliance, the Soviets and the Nationalists had in fact helped one another for years against the Japanese. The existence of Chiang's army to the south, and the forty divisions of the Red Army to the north, had each helped keep the Japanese Kwantung Army moored in Manchuria, unavailable either to fully assist in the conquest of the rest of China, or to attack into Siberia (Lyons).

In the treaty Stalin sought with Chiang Kai-shek, the Soviet Union would formally recognise Chiang's Nationalists as the government of China and would aid them in their fight against the Japanese, while withholding aid for that fight from the Chinese Communists. In return, Stalin sought Chiang's recognition of the Mongolian People's Republic (a Soviet ally) as an independent nation, along with concessions in Manchuria regarding Soviet control of the ports of Dalien and Port Arthur, and the Manchurian Railway, which connected to the Trans-Siberian Railway (Chandler et al.) Such a treaty would secure the long-term political, military and economic gains Stalin sought from the Chinese leader. To obtain them, he was willing to forego helping his supposed communist comrades-in-arms. At the time, Stalin did not give Mao Zedong's rural peasant communist movement much chance of success. At best, he thought they might someday earn a place in a weak coalition Chinese government, over which the Soviets could exercise considerable sway (R. Bernstein).

Chapter 7

The *Enola Gay* and *Bockscar*

By 2 February 1945 there had been no official reply from the Germans or from the Japanese to the January ultimatum. Roosevelt therefore authorised the Army Air Force to proceed with the initial bombings as soon as weather conditions were favourable in both locations. The theatre commanders Eisenhower, MacArthur and Nimitz had been briefed on the imminent atomic missions and directed to continue with their land, air and sea operations as they had planned. The Pentagon chiefs hoped these new weapons might themselves force their enemies' prompt surrender, but no one could be sure of that. Any diminution of effort by the Allies could be mistakenly interpreted as weakness or exhaustion. So the atomic bombing of a German city would occur on the same day as Allied bombers rained conventional explosives on perhaps several others, and while Eisenhower's armies continued their eastward advance into Germany. Similarly, the atomic destruction of a Japanese city would occur on the same day as Curtis LeMay's B-29s raided several others, and air raids preparing for troop landings on Japanese-held Iwo Jima would continue as they had for several months.

The 509th Composite Group's two squadrons had deployed in late 1944 eastward and westward, to a bomber base at Rattlesden, Suffolk, England, and to Tinian Island in the Marianas. On 6 February 1945, the B-29 *Enola Gay* piloted by Colonel Paul Tibbets dropped a plutonium bomb on Yokohama, Japan at approximately 4 p.m. local time (Map 2, page 48).

At approximately 8 a.m. local time, the B-29 *Bockscar* piloted by Major Charles Sweeney dropped a uranium bomb on Dresden, Germany (Map 3, page 49). Thus the attacks were virtually simultaneous, and marked the first and only time during the war that specific actions were synchronised between the Pacific and the European theatres. Since the bombardier on the *Bockscar* was RAF Wing Commander Paul Barnstable, Great Britain was clearly a partner in the initiation of the age of atomic warfare.

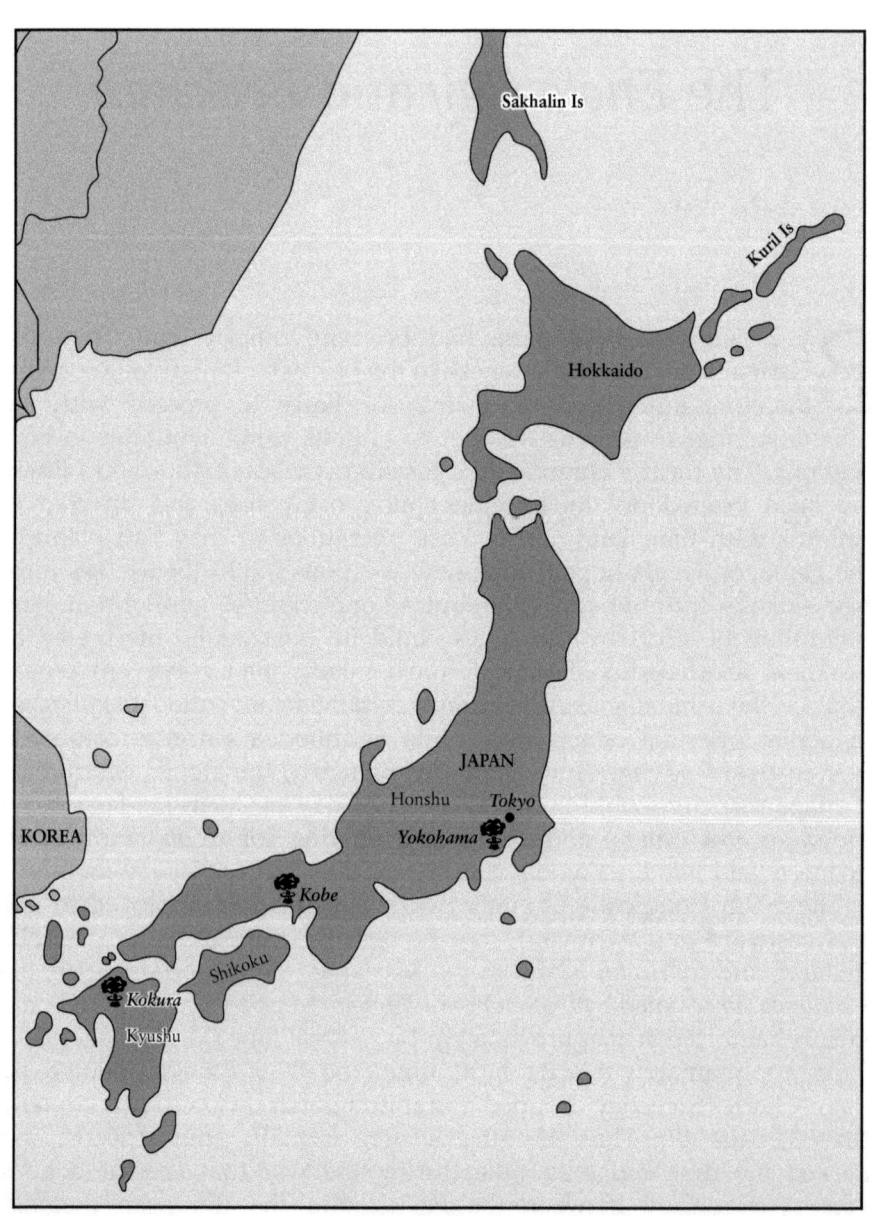

Map 2: Japan 1945

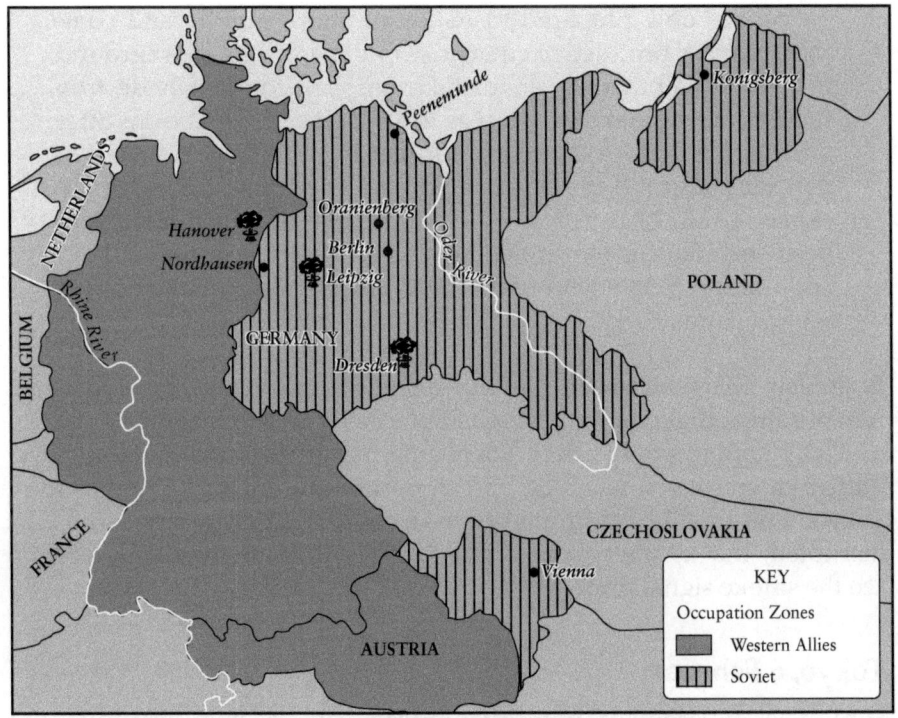

MAP 3: Germany at the End of the War

As soon as the Pentagon confirmed that both planes had accomplished their missions, the White House issued a press release revealing the simultaneous destruction of military assets 6,000 miles apart, by the use in each case of single bombs with the power of over 10,000 tons of TNT.

> 'With this bomb' the President's announcement said, 'we have now added a new and revolutionary increase in destruction to supplement the growing power of our armed forces. These bombs now in production are atomic bombs, harnessing the force from which the sun draws its power. Before 1939 it was the accepted belief of scientists that it was theoretically possible to release atomic energy, but no one knew any practical method of doing it. By 1942 however, we knew that the Germans were working feverishly to add atomic energy to the other engines of war with which they hoped to enslave the world. But they failed. We have now won the battle of the laboratories.

'We are now prepared to obliterate more rapidly and completely every productive enterprise our enemy has aboveground in any city. Let there be no mistake; we shall completely destroy their power to make war. It was to spare their people from utter destruction that we issued our ultimatum. Their leaders rejected that ultimatum. If they do not now accept our terms they may expect a rain of ruin from the air the like of which has never been seen on this earth. Behind this air attack will follow land and sea forces in such number and power as they have not yet seen.' (Truman 1945)

Roosevelt, Marshall and a handful of others knew that the threatened rain of ruin at that moment consisted of exactly six more bombs.

The planning group chose Yokohama in part because of its military-industrial and naval facilities, and in part because it was 20 miles from Tokyo. Presumably communications from the stricken city would be disrupted, but as one planner put it: 'Ol' Hirohito should be able to see the smoke signal pretty well.'[1]

Tokyo, 6 February

Hirohito's government ministers indeed saw it, and heard it, and did not believe it. There were no communications with Yokohama. Pilots very soon brought back reports of such widespread destruction that their commanders refused to believe them. In the fading light of evening, the commanders insisted, the pilots had misinterpreted what they had seen. Soon, Japanese broadcaster NHK was receiving the text of FDR's announcement, and was asking the Imperial General Staff what an atomic bomb was.

Most staffers had no idea, but a few did. A few senior officers knew that Japan had its own programme to develop atomic weapons. But it consisted of hundreds of people, with a budget roughly equal to about $100,000, and it so far had produced miniscule quantities of uranium useful to fuel a bomb (Dahl).

The Navy contacted one of the academics involved in that programme, Professor Asada, and had him fly as an observer over Yokohama at first light next morning. The skies were filled with smoke from the fires that had raged through most of the city all night. Asada had

---

1. See 'What Really Happened' (6 Aug 1945).

himself seen the single massive mushroom cloud from his office window in Tokyo. He understood that this had not been the work of thousands of bombs, but of one, and that only nuclear fission could have powered it.

The day of the first atomic bombing Navy Minister Yonai wrote, 'The war is lost' in a top secret memo referring to the destruction wrought by the bomb (Brooks). But the Army's Air Chief, General Anami declared 'I am convinced that the Americans had only one bomb' (Gray). (The Japanese had not yet confirmed the atomic destruction of Dresden.) Then Professor Asada reported to Naval Headquarters that the Americans had indeed used an atom bomb, contrary to the consensus of a 1943 committee of Japan's leading physicists that the enormity of the effort to develop an atomic bomb and acquire the enriched uranium to fuel it was too great for even the US to do so 'before the end of this war' (Dahl). Still clinging to this position despite the evidence he'd just seen with his own eyes, Asada told the military that the process of making such a weapon would have to have been so enormous an undertaking that it was inconceivable that the Americans could have built more than five or six of them (Brooks). As discussed above, he was just about right on that last point, but they were six more atom bombs than anyone else in the world had.

Berlin, 6 February

Six thousand miles to the west, Dresden was a city of considerable cultural significance. But it was also home to hundreds of factories producing materiel for the German war machine (Angell). It was a major north–south and east–west road and rail junction important in the movement of material and troops (F. Taylor). It was also far enough east, and therefore close enough to the advancing Red Army that bombing Dresden would cause damage and logistical disruption to the German forces trying to resist the Russian advance. The Soviets had asked the Western Allies to provide what assistance they could (Keegan) and especially to disrupt the Wehrmacht's shifting of forces toward the Russian front (Davis). Based on these considerations, the Western Allies had developed a plan in January 1945 for a major bombing mission in mid-February on Dresden. That plan was to involve hundreds of bombers, in multiple waves, dropping a mix of high explosive and incendiary munitions with the intent of causing maximum blast and fire damage (Longmate) in a city with many timbered buildings (de Bruhl).

This firebombing plan, similar to raids that caused firestorms in Hamburg and other cities earlier in the war, was superseded by the

> ### Japan's Own Bomb?
>
> In 1946 an article in the *Atlanta Constitution* claimed that a Japanese military officer had told of witnessing an atomic bomb test in Japanese-held Korea in the last days of the war. According to the article, the well-placed informant said that this super-secret Japanese military programme had been forced to relocate to Korea as the USAAF's bombing of Japan intensified. 'We lost three months in the move. If it hadn't been for that, we would have had multiple *genzai bakuden* (atomic bombs) loaded on kamikaze aircraft' the informant is quoted as asserting (Snell). The article provided no corroborating evidence, but speculated that the several days' delay in the Japanese government's acceptance of the Allies' call for surrender was motivated by Tokyo's hope to be able to deploy their own atomic arsenal. The lack of supporting evidence has not prevented this notion from resurfacing periodically even to the present day (Benke; Wilcox). For some, 'evidence' is an unnecessary frill.

decision to make Dresden the first atomic target in Germany. Dresden's location 100 miles south of Berlin, and 60 miles from the Russian front would help ensure that both German and Russian leaders understood the significance of the city's destruction.

Nazi Propaganda Minister Joseph Goebbels is reported to have wept with rage for twenty minutes after he heard the news of the destruction of Dresden. He then launched into a bitter attack on Hermann Göring, the commander of the Luftwaffe, the German Air Force: 'If I had the power I would drag this cowardly good-for-nothing before a court. . . . How much guilt does this parasite not bear for all this, which we owe to his indolence and love of his own comforts . . . ?' (Reiman). Goebbels promptly issued a press release decrying the Allied attack on Dresden, stating that it had no war industries and was a city of culture. He quickly followed with a leaflet showing burned children and the claim that 200,000 people, mostly refugees, had been killed in the attack (F. Taylor).[2]

The Western Allies

There may have been some who felt sorry that the Nazi homeland had been so brutally attacked, but such sympathisers were not to be

---

2. See 'What Really Happened' (13 Feb 1945).

> **6 FEBRUARY 1945**
>
> In 1946, a Pulitzer prize-winning war correspondent who had covered the war in the Pacific and in Europe authored a book that won critical acclaim. In restrained prose, the book traced the lives of a series of atomic bomb survivors. Albert Einstein was said to have ordered a thousand copies for distribution and a great many newspapers clamoured to serialize it (Severo). John Hersey had visited both Yokohama and Dresden in researching the parallel stories of the two cities destroyed on the same day in his simply titled *February Sixth*.
> (See 'What Really Happened' 31 Aug 1946)

found in Coventry in England, which the Nazis had systematically firebombed, nor elsewhere in Great Britain, which had endured so many air raids, and V-1 and V-2 missile attacks. Nor were there likely many sympathisers in any of the formerly occupied European lands. Nor even in America, where outrage was growing as more accounts emerged about the numerous and appalling German atrocities.

Well over a hundred thousand 'Japs and Krauts' as many GIs would call them, died on 6 February 1945, and the world waited to see if the Axis governments would recognise reality, or would challenge the Allies to continue the rain of ruin. While the White House listened through diplomatic channels, the Pentagon directed the 509th Composite Bomber Group to prepare to deliver two more atomic bombs on 9 February, weather permitting, and unless halted by the President. Eisenhower's four million troops from the US, UK, Commonwealth, French and other nations continued to press eastward on a front from the North Sea to the Swiss border (Eisenhower 1948). For example, the American Seventh Army had reached the heavily fortified Siegfried Line, also known as the Westwall, on 1 February, and breached it on 9 February (Keegan).

In the Philippines, MacArthur's forces had begun the Battle of Manila on 3 February and it continued with fierce urban fighting (Connaughton et al). Fifteen hundred miles to the north, the US Navy continued preparations for the invasion of Iwo Jima (Morison).

Chapter 8

Soviet Advances

For their part, the Russians added to the pressure on the Axis both in the east and the west.

Europe

As noted previously, Stalin was intent on getting to Berlin before the Western Allies. In late January he had approved his front commanders' plans to resume their westward advance after taking less than a week to resupply their two million troops roughly at the line of the Oder River. Marshall Zhukov's lead elements were less than 40 miles from Berlin, and Konev to his south was behind, but not by much. The presumption in the Kremlin was that Zhukov would reach Berlin first, and take it, but Konev's westward push would put his troops in position to swing north/north-west and potentially get to Berlin at about the same time (Lunde). Stalin's risk-laden decision about whether to delay his army's drive toward Berlin in order to protect its northern flank has been much studied and discussed to the present day (Glantz 1982).

There were still major, potentially threatening German formations to the north of the Soviet Fronts, but Stalin sensed there was a brief opportunity for his forces to capture Berlin before either Eisenhower's armies did so, or before the Americans' new atomic weapons perhaps ended the war.

Though they were not fully refitted and resupplied from their January advance across Poland from the Vistula, Zhukov's and Konev's armies were nonetheless in far better shape than the much-depleted units of the German Wehrmacht (the Regular Army) and Waffen-SS (military units specifically formed as part of the Nazi Party) which they faced. Some units consisted of naval and air force personnel summarily pressed into service as infantry. Others consisted of Hitler Youth, boys under military age nonetheless thrown into combat operations. There were also newly mobilised citizen militias of the *Volkssturm* (Weinberg; Lunde).

Numbers mattered too: even after assigning eight divisions to protect his northern flank and prevent the Germans from cutting in behind him, Zhukov's million-man Front still included dozens of divisions for his drive on Berlin. Konev's Front was of similar size. Facing these Soviet forces were perhaps 400,000 Germans (Lunde).

Marshal Zhukov, starting from positions at the Oder River about 40 miles east of Berlin, attacked the German defences along the Seelow Heights west of the Oder on 4 February. Although the 300ft 'heights' were not in themselves too formidable, the marshy ground between the heights and the Oder proved unexpectedly difficult. An early February thaw melted the snow from a 27 January blizzard and turned much of the whole region to mud (McTaggart). Zhukov's lead elements took longer than planned to negotiate the 6–8-mile wide quagmire in front of the Seelow Heights. Instead of taking one day as planned, it took about three.

Stalin was annoyed by Zhukov's slow progress; on the second day of the advance from the Oder, he authorised Koniev, to Zhukov's south, to redirect a portion of his westward-moving Front to swing north-west and approach Berlin from the south.[1]

The mud that slowed Zhukov down also helped protect his northern flank. When Zhukov began his push toward Berlin, General Rokossovsky's 2nd Belorussian Front on the north had not advanced as far west as Zhukov, and had not yet neutralised the thirty (severely depleted) German divisions that remained in Pomerania and Prussia.

On 8 February Army Group Vistula under never-before-in-combat Heinrich Himmler launched a hastily-organised southward counter-attack from Stargard. It had been moved up from its original planned launch of 15 February due to the strongly renewed westward push by Zhukov. Several German divisions made some miles of initial progress. But lack of effective leadership, muddy conditions and prompt pushback from the Soviet forces that Zhukov had left on his northern flank stalled the German advance. It was called off after just three days (Le Tissier; Duffy).

This not wholly surprising German counter-attack might have given Stalin some pause. He might have decided to halt Zhukov's westward advance, directing him instead to attack northward and work with Rokossovsky to 'clean out' the Germans to his north. But the prize

---

1. See 'What Really Happened' (16 Apr 1945).

of Berlin beckoned, and, knowing from his spies that the Americans had the atom bomb, he did not know how much longer he had to try to capture Hitler and his capital. Stalin did not let Himmler's feeble pressure from the north slow the Red Army's advance on Berlin[2] (Map 1, page 40).

The Battle of Stargard had one unexpected consequence: after the Soviets' successful counter-attack against the Nazis' attempted advance out of Stargard, divisions of the 61st Army of Zhukov's 1st Belorussian Front were then positioned less than 150 miles from the Germans' rocket research and development facility at Peenemünde.

To Zhukov's south, Marshal Konev's armies also rapidly swept toward Berlin in a broad arc from the south-east. On 8 February, two days after Dresden was destroyed, and Stalin thereby had confirmation of the Americans' ability to use the new weapons, the Soviet high command sent Eisenhower's headquarters a message. It told him that Red Army troops expected to be in Berlin very soon. 'Comrade Stalin' the message read, 'would very much appreciate it if the US Army Air Force did not destroy Berlin with Red Army troops in it.' The message was of both tactical and geopolitical significance, for Stalin was also thereby staking claim to the 'prize' of taking Berlin, earned by the blood and sacrifice of millions of Soviet soldiers. Red Army units reached the eastern portions of Berlin on 9 February and began firing artillery into the city.

The Soviet Far East

Before dawn on 8 February (two days after the destruction of Yokohama) the Soviet Far Eastern Command under Marshal A.M. Vasilevsky struck across the Manchuria border. This came about two hours after the Soviet Foreign Minister in Moscow handed a declaration of war to the Japanese Ambassador. In the previous four weeks, Vasilevsky had covertly thinned out the border defences along much of the 2,600-mile Soviet–Manchurian border. At this point in the war, a Japanese offensive into Siberia was a negligible risk. Cavalry units from the Soviets' Mongolian allies helped maintain the border defences and ensure that the Japanese were not aware of the movement of Soviet divisions behind those lines. Preparations were kept far back from the border to help maintain secrecy. Soviet doctrine considered surprise to be the key to victory (Glantz 1983), and it would be especially critical in this rapidly-planned operation.

---

2. See 'What Really Happened' (15 Feb 1945).

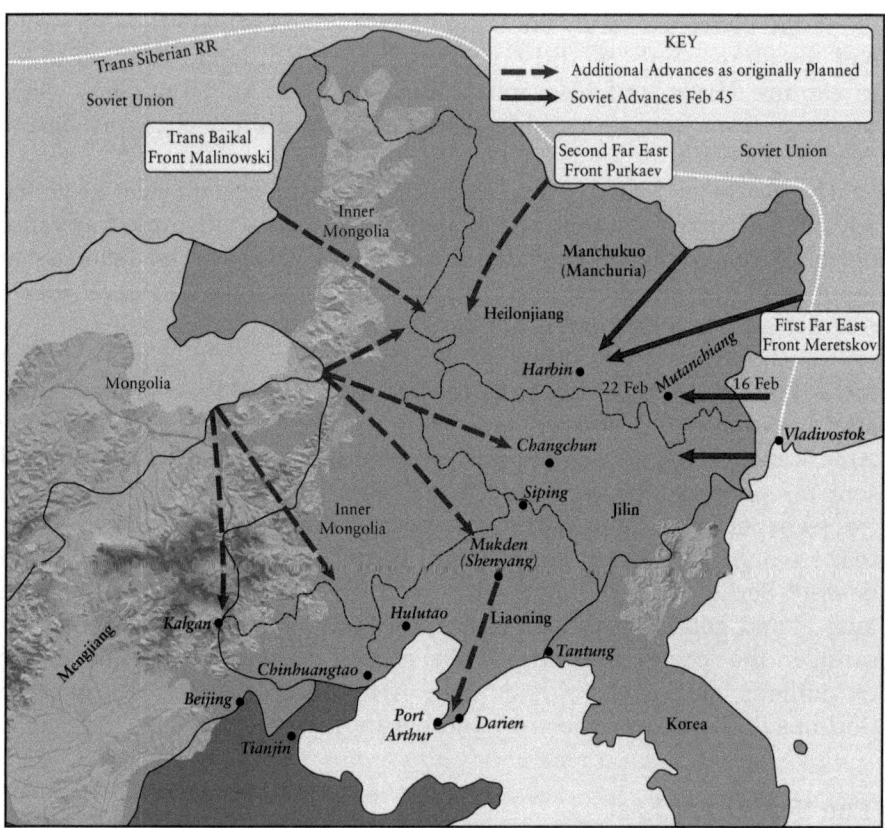

MAP 4: Soviet Manchurian Campaign: Plans and Execution

Supplemented by a few rear-area divisions Stalin had shipped out on the Trans-Siberian Railway, the Soviet Far Eastern Command was able to attack Manchuria with about 35 divisions (about 425,000 troops). These were not the Red Army's most experienced, nor best trained or equipped. But neither were they facing Japan's finest. The once-elite Kwantung Army holding Manchuria had been reduced to about 700,000 troops, many under-trained and under-equipped. Many of the best Kwantung Army units had been redeployed to the Pacific to try to hold onto islands like Saipan (Hornfischer). In China there were also about 200,000 Chinese troops of the puppet regimes of Inner Mongolia (the Menjiang United Autonomous Government) and of Manchukuo. All were focused on trying to defend the interior of Manchuria rather than maintaining a strong posture all along the border (Glantz 1983). The Japanese assumed that the Soviets would overstretch their logistical capabilities if they tried to

SOVIET ADVANCES

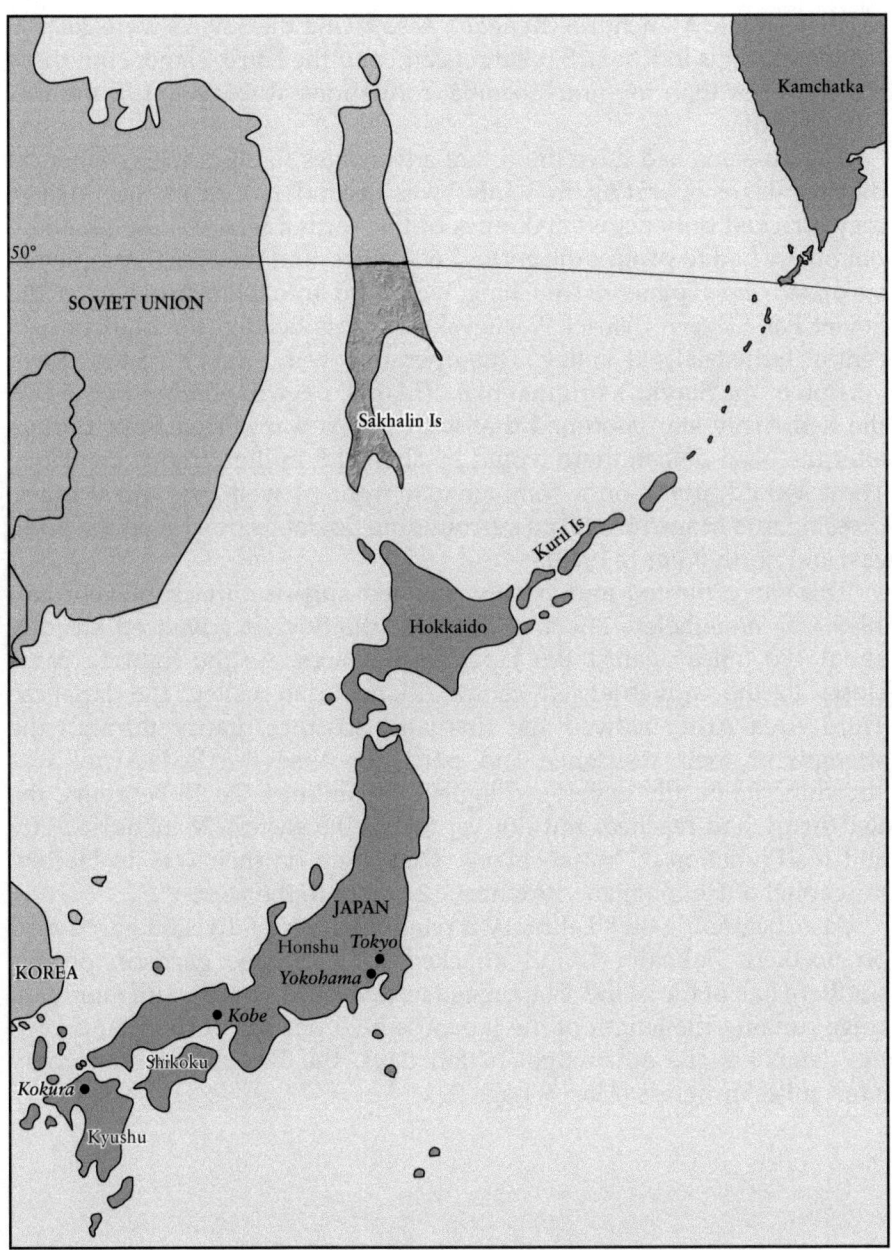

MAP 5: Sakhalin and Kuril Islands

reach deep into Manchuria (Keegan). Also facing the Soviets were 280,000 Japanese troops in Korea, Sakhalin Island and the Kuril Islands, but these too were less than top-notch combat formations at this point in the war (Hasegawa).

The Russians had three important advantages to offset their numerical disadvantage: operating in winter was second nature to the Soviets; they attacked only across 400 miles of the border, whereas the Japanese nominally had to protect the entire 2,600 miles; and the attack completely surprised the Japanese. Attacking westward into Manchuria from the Soviet Far Eastern district, Vasilevsky's forces headed for Manchuria's central industrialised valley. This operation was a much scaled down version of the Stavka's original plan. (Map 4, page 58) Earlier in the war the Red Army staff assumed that with troops transferred from Europe after the Nazi defeat, there would be about 1.5 million troops available. These would attack on a semi-circular front of well over 1,000 miles, pressing into Manchuria from surrounding Soviet territories on the west, east and north (Glantz 1983).

This more limited and hastily planned surprise attack of February 1945 was nonetheless effective. Initially, the Soviets advanced steadily about 100 miles against the Japanese defences. As the fighting came closer to the industrialised central Manchurian valley, the Japanese Third Area Army slowed the Russians' advance, partly through the strength of their resistance and partly because the Red Army was stretching its hastily developed logistic capabilities. On 16 February, the Red Army had reached, but not yet taken, the eastern Manchurian city and road junction of Mutanchiang. They were on their way to Harbin, the capital of Heilongjiang Province, 120 miles to the west.

Also, beginning on 8 February, a reinforced Soviet division garrisoned on northern Sakhalin Island attacked the Japanese garrison on the southern half of the island. The rugged terrain of the north-south mountain ridge running the length of the Japanese-held portion of the island gave the defenders the advantage. In ten days, the Red Army made only a few miles' progress (Map 5, page 59).

Chapter 9

Last Days

Japan, 9 February

Prime Minister Kuniaki Koiso convened a meeting of the Supreme Council for the Direction of the War. This was 'The Big Six' (Admiral Yonai, Navy Minister; Army General Anami; General Yoshijirō Umezu, Army chief; Admiral Koshirō Oikawa, Navy chief; and Mamoru Shigemitsu, Foreign Minister). Reports of Russian advances in Manchuria and on Sakhalin Island were just beginning to come in when messages also arrived announcing another atomic bombing. This time, the *Straight Flush* piloted by Captain Claude Eatherly out of Tinian had delivered its device over the large port city of Kobe (Map 2, page 48). Like Dresden, Kobe had been slated for an incendiary attack in early February (Craven and Cate), but that plan had been superseded by this 'more efficient' attack plan.[1]

The Japanese leaders were still processing the information they were receiving about what used to be Yokohama. Advisors told them that the carnage and destruction of the atom bomb far exceeded the devastation caused in that port city by the Great Kanto Earthquake of 1923. There were many tens of thousands of casualties, reportedly including horrific burns and frightening sicknesses. What barbarity the Japanese people would have to endure, the government leaders thought, before the Japanese Empire forced the Americans to accept the Japanese terms for ending the war! Now, Kobe was added to the list of outrages that would only stiffen the people's will. Or so the hard-line members tried to convince each other. Now to the savagery of the Americans was added the treacherous surprise attack by the Russians. How dare these barbarians promise neutrality and then launch a surprise attack.

---

1. See 'What Really Happened' (9 Aug 1945).

(Of course, when the samurai of old launched surprise attacks, or the modern Japanese military did the same against the Chinese in 1895, the Russians in 1904, or the Americans in 1941, that was just using an established pillar of the noble warrior tradition [Hoyt].)

The War Council could not agree with Foreign Minister Shigemitsu's proposal to approach the Allies through neutral channels to seek a negotiated peace. There had been low-profile peace factions in various parts of the government since 1944 (L. Brooks). In August 1944 Admiral Yonai had tasked Admiral Sokichi Tagaki to assess peace prospects (Hasegawa) after Tagaki's research of combat and domestic data had convinced him that Japanese defeat was inevitable (Butow). Yonai knew that the Imperial Japanese Navy was becoming a thing of the past, as was made stunningly clear in October 1944 by the loss of twenty-six major ships, including four carriers, three battleships, ten cruisers and nine destroyers at the Battle of Leyte Gulf (Morison). Yonai was also aware that Allied submarine warfare and aerial minelaying had, by January of 1945, essentially destroyed the Japanese merchant fleet, so essential for importing the food, fuel and other resources the Home Islands needed (Blair). Now, after the destruction of two Japanese cities by atomic bombs, Yonai supported Shigemetsu's idea of seeking peace terms.

Berlin, 10 February

There was no such War Council in Berlin. The war was run hour by hour by Hitler, ensconced now full time in his well-fortified bunker 30ft under the Reich Chancellery garden. In the Fuhrer's megalomaniacal system, there were no consultative councils, and those around him disagreed with him at their peril. Nevertheless, on 30 January 1945 Armaments Minister Albert Speer wrote to Hitler that 'the war is lost', citing the loss of 60 per cent of Germany's coal supply in territory taken by the Red Army. Coal was by then essential not only for heat and power, but for synthetic oil and motor fuel (Read and Fisher).

That same day too, there was a riot in Berlin over sparse food supplies (Read and Fisher). As a US Army historian wrote:

> For the German people, the first week of February was the darkest of the war ... Three weeks earlier; the front had still been deep in Poland and nowhere on German soil. Now Upper Silesia was lost; in East Prussia a German army group was being cut to pieces; West Prussia and Pomerania were being defended by a skeleton

army group under a novice commander; and the defence of the Oder would have to be entrusted to armies that had already been defeated on the Vistula and chased across Poland. If the Russians maintained their rate of advance, and there seemed to be no reason why they could not, they would be on the Rhine in another three weeks (Ziemke).

Hitler was not only physically but psychically walled off from the people of Germany. The news of the destruction of Dresden elicited little more than a shrug and a wave of the hand. 'What is the situation in East Prussia?' he wanted to know. Reports that tens of thousands of German civilians had died instantly in Dresden made essentially no impression on The Leader of the German People. The reality above ground was horrific. Yet in the subterranean land of fantasy, Hitler confidently positioned non-existent formations for decisive battles with the very real Russian armies on the outskirts of Berlin. So the atomic destruction of a second German city, Hanover, on 10 February made little noticeable difference to the manic proceedings in the *Führer's* bunker.

Germany, 9–19 February

Having received no capitulation message from either Axis capital, the White House directed the Pentagon to proceed as planned: two more weapons were to be dropped on the third day after the first pair. Accordingly, as noted, Kobe was destroyed on the 9th, but northern Europe's weather delayed parallel operations over Germany that day. Instead, on 10 February 1945, the *Full House*, piloted by Captain Rick Taylor operating out of Rattlesden airbase in Suffolk, dropped an atomic bomb onto Hanover, Germany. An important transportation and industrial centre, Hanover was slated to be within the British zone of occupation.

The loss of tens of thousands more German civilians made far less impression on Hitler than the frenzied reports of the encirclement of Berlin by the Red Army. Marshal Konev's troops, advancing from the south-east linked up near Potsdam, west of Berlin, with Marshal Zhukov's troops, who had advanced around the north of Berlin (Lyons) (Map 1 page 40). It was 12 February; Hitler had refused to leave the city earlier; now he could not. He continued to order the city's defenders to hold out until they were relieved by the several huge highly capable units that existed only in his mind. Hour by hour, Zhukov's and Konev's forces fought their way block by block toward the Chancellery.

ENDING THE WAR

In the last days of the Reich there were peace feelers from several different Nazi sources. Among these were Foreign Minister Joachim von Ribbentrop and Reichsführer-SS Heinrich Himmler (Read and Fisher). Himmler, encouraged by Heinz Guderian, made indirect contact with Count Bernadotte, head of the Swedish Red Cross, to explore a German surrender to the Western Allies for the purpose of joining forces with them against the Soviets (Padfield). Eisenhower's headquarters would accept nothing less than an unconditional surrender, and word of the attempted negotiations was soon broadcast by the BBC. When this information filtered into Hitler's bunker, the *Führer* was enraged and dismissed Himmler from all command (Kershaw).

Meanwhile, Hermann Göring, ensconced at his estate in Obersalzberg, was persuaded by Generaloberst Alfred Jodl and others that Hitler's decision to remain and die in Berlin necessitated that Göring take power as Hitler's designated successor (Manvell). When Göring sent a message to this effect to Berlin, Hitler's aide Martin Bormann persuaded Hitler that Göring was a traitor. Himmler and Göring, two of Hitler's most trusted accomplices and most likely successors, had abandoned him. As he dictated his last will, he named Grand Admiral Karl Dönitz to succeed him as President of Germany (Kershaw).

Zhukov and Konev's troops owned more and more of Berlin, but the fighting was block by block, building by building. For those in the bunker, there would be no escape. Nor was there escape for the citizens of Leipzig, who died from the third atomic weapon dropped on German soil on 14 February. Besides destroying many thousands of civilian lives, the blast also disrupted movement of German troops seeking to attack the Russians besieging Berlin, 80 miles to the north.

Military professionals such as Jodl and Guderian got FDR's message: they were experiencing the rain of ruin he had promised. Every few days, a single bomb would destroy a German city more thoroughly than huge fleets of USAAF and RAF bombers had ever done. But Hitler dismissed Guderian as Army Chief for his defeatist attitude (Chen). *Der Führer* would allow no discussion of defeat, even as he could feel the rumble of Soviet artillery fire in the concrete walls of the bunker. He continued to issue orders to his field units, only a few of which existed.

By 18 February, it was clear that no one was going to prevent the Red Army from reaching the Reich Chancellery. Hitler killed himself that day. His designated successor Admiral Dönitz announced the death of Hitler to the German nation (Shirer). In that same address, on 19 February, he announced the unconditional surrender of all

German military forces. The German war was officially over as of 20 February.[2]

Thus the most violent conflict in European history ended with a final spasm of horrific violence. Millions of Allied troops were relieved. Many of their commanders were surprised, but shifted swiftly to thinking about the post-war occupation. On 6 February the Allies had agreed to the occupation zones for the Reich (Mosely). Eisenhower's headquarters had a plan, Operation Eclipse, for the conduct of the occupation (McCreedy). As the Eclipse plan was launched, American, British and other Allied units advanced eastward to the limits of their assigned occupation zones.

Tokyo, 10–18 February

Even after the second atomic bombing of Japan, the War Council was still divided. General Anami maintained that the bombing was not important (L. Brooks). Others maintained that world opinion would not stand for continued US use of atomic weapons (L. Brooks). There was also concern that the millions of Japanese troops in the field would find the idea of surrender too alien to obey (L. Brooks).

The Emperor's advisor, Koichi Kido, conveyed the Emperor's observation that 'We must bow to the inevitable. We must put an end to the war as speedily as possible' (L. Brooks). The War Council and the full Cabinet met to grapple with how that might be done. Despite the Allied ultimatum's clear demand for unconditional surrender, the Japanese ministers and military professionals nonetheless sought to demand conditions such as self-disarmament, no occupation and Japanese control of war crimes trials (Frank).

On 11 February, Japanese radio network NHK picked up broadcasts announcing that on the previous day another atomic bomb had fallen in Germany, on Hanover. That was four bombs so far. The US was clearly not hesitating to continue using them. How many more did they have? Navy advisor Professor Asada had estimated the Americans could only have produced about five or six. If he was right, the rain of ruin might stop soon, at least for a while. But did Asada know what he was talking about? Two years previous he'd been part of the committee of physicists who had asserted that the US could not possibly develop such a bomb by this time.

---

2. See 'What Really Happened' (30 Apr–7 May 1945).

Moreover, there were also the conventional bombs to consider. The seemingly unlimited numbers of the new B-29 bombers flying out of Tinian, Saipan and Guam could now reach virtually all of Japan. The pace of destruction from mass air raids was increasing, especially since the beginning of the year (Craven and Cate).

Then there was the external and the internal threat from the communists. The Soviet Red Army was visibly advancing in Manchuria. At home, there was a worrisome Japanese communist movement. Early in February, Prince Fumimaro Konoe warned the Emperor that a communist revolution was as much a threat to him and the Imperial System as was a US victory (Hasegawa).

Anami nevertheless argued that the mighty Japanese Army could still extract enormous costs in Allied lives, costs so high that the Allies would accept Japanese terms. He cited the imminent battle for Iwo Jima. 'They have been preparing to invade for months. But they have hardly imposed a scratch on the defenders, and when the Americans hit the beaches, we will turn the ocean red with their blood' he boasted. 'The Americans are weak – they believe each of their miserable people's lives matter.'

'"Blood-red beaches". Really?' asked Shigemitsu (although not quite so bluntly, given the typically circumlocutory habits of Japanese,) 'What makes you think they will risk even a company of their Marines when they can just blast the whole miserable island with one of these bombs? Next they'll probably go for Okinawa, and they'll take it with a hundred thousand Japanese deaths and maybe one GI, but only if he falls asleep and drives his jeep over a cliff inspecting the bomb damage.'

And so the arguments went on, as Yokohama and Kobe still smouldered, and thousands of citizens showed up at emergency clinics with the mysterious deadly disease today known as radiation poisoning.

In the face of the continuing atomic destruction, the prolonged machinations of the Supreme Council and Cabinet might seem to be inexplicable behaviour for politicians. But these were not politicians in the Western sense of men whose power and position depended on votes of the citizenry. Japan in the previous hundred years had adopted many of the trappings of Western culture. But like the formal attire Japanese leaders were often photographed wearing, this was largely superficial; deeper down, the ministers were more like rival samurai clan leaders, oligarchs jockeying and protecting their positions through the exercise of power, not electoral popularity (Hoyt). They were supposedly subservient to the Emperor, but as the Army's independent action in taking over Manchuria in the 1930s showed, the samurai did not always allow the

Emperor the power which the Meiji Restoration had theoretically given him (Hoyt).

Nevertheless, late on 12 February, the ministers agreed to take the highly unusual step of asking for an Imperial Conference, and seeking the Emperor's view. The meeting convened on the morning of 13 February. Anami argued for the hard line, for giving however many more citizens the honour of dying in the hope of 'forcing' the Americans to grant the terms Japan demanded.

Several of the military leaders had been touting the effectiveness of the new kamikaze ('Divine Wind') suicide campaign which had debuted in October 1944 during the Battle of Leyte Gulf. (But as noted, that battle hadn't ended well for the Japanese. Suicide pilots had sunk a US carrier and damaged several other ships by crashing their planes onto the ships [Toland].) As of January 1945, specially designated kamikaze units, with minimally-trained pilots and relatively flimsy purpose-built aircraft were readying for action against the expected Allied invasions of Japanese-held islands. 'With these tactics' it was said, 'our noble warrior pilots will inflict more damage on the invaders' fleets than they can bear. Our total ruin will be prevented!'

It was true that the initial kamikaze attacks had taken the Allied navies by surprise and had caused some damage. We cannot know how much more damage more intense suicide flights might have caused to the ever-larger fleets planned for the invasions later in 1945 of Iwo Jima, Okinawa, and ultimately the Home Islands. There is one estimate of the number of ships lost to kamikaze action during those 1945 invasions exceeding the losses inflicted by the entire Japanese fleet in all the previous years of the war (Hastings 2009b).[3] But, as perhaps the Japanese admirals and generals themselves realised, even sinking dozens of US ships in 1945 would have had little effect on a navy that by then had 100 aircraft carriers among almost 900 major surface warships (Naval History and Heritage Command).

At any rate, the future potential of the kamikaze campaign could not compare to the real present effects of the atom bombs. A squadron of Japanese pilots might sink part of an invasion fleet, but now a single American pilot could wipe out a city. Technology was vanquishing valour. Westerners might call it technological determinism (Hastings 2012). Or, as Admiral Yonai put it, the atom bombs were like 'gifts of

---

3. See 'What Really Happened' (Feb–Mar 1945(2)).

the gods' (Buruma) which made it possible for the Japanese military to surrender in the face of science, not of superior fighting spirit (Frank).

Then the conferees received the news – that morning, 13 February, Kokura, an industrial centre and arsenal, became the third Japanese city to turn into a radioactive ruin. Most accounts, such as from the Emperor's advisor Kido, report that Hirohito was already determined in his view before the news of this additional atomic bombing reached him. The Emperor told the conference that he wanted the Allied ultimatum to be accepted right away, with the sole condition that the Imperial prerogatives would be preserved (Frank). Still, the Cabinet dithered and stalled. Finally that evening, the Foreign Minister eluded the military opponents and got a message broadcast on the civilian radio that evening. It announced the acceptance of the Allied ultimatum, provided that the Emperor's prerogatives as sovereign ruler were preserved (L. Brooks).

Welcome as the message of 13 February was in Washington, London and Moscow, it was not enough. To be sure, the US and UK believed they would indeed have to preserve 'the Imperial System' at least long enough to secure the surrender of all of the far-flung Imperial military forces, and to ensure a level of civil peace as the occupation took hold (Frank). But neither FDR nor Churchill was prepared to let Hirohito retain all the Imperial prerogatives that had led his nation into this war. Nor could the Allies appear to be bowing to such an Imperial demand.

Therefore, two days later, with her Allies' concurrence, the US replied to the Japanese, indicating that the Japanese people could ultimately choose their form of government but that the Emperor would be subject to the Supreme Commander of the Allied Powers (L. Brooks). This reply caused further debate in Tokyo when it arrived on the 15th. There were rumours and threats of coups. Lieutenant General Okido was chief of the military/secret police, the Kempeitai. He told the cabinet secretary: 'If Japan surrenders, the Army will rise. Tens of millions of lives may be sacrificed, but we must not surrender' (L. Brooks). There were posters on the streets and the subway branding the peace faction as traitors (L. Brooks). On the 15th also came word that the Red Army was nearing Mutanchiang in Manchuria and was poised to then advance on the major city of Harbin.

The debate now centred on the fate of the Emperor personally, and of the paramount importance of preserving the Imperial System as a whole (i.e. the system by which these modern samurai chiefs jockeyed with one another for greater or lesser shares of the true power in the state). Some argued that the wording of the Allied response clearly indicated that the Allies had no intention of eliminating the Emperor.

In keeping with the US plan to continue the 'rain of ruin' every few days, a fourth atomic bombing mission was slated for the 16th. The President put it on hold, but directed that the USAAF be ready to carry it out if he felt that the Japanese were stalling. He wanted the war over as soon as possible, largely for the sake of US troops, but also because every day of continued fighting gave the Soviets more claim to a piece of the victory. If necessary to force a Japanese capitulation once and for all, Roosevelt was prepared to annihilate yet another Japanese city.

The 'atomic hold' did not extend to a conventional raid by the US Navy. On the 15th, the sixteen aircraft carriers plus several battleships, cruisers and dozens of destroyers of Task Force 58 was about 100 miles off the Japanese coast, near Tokyo. They launched the better part of 1,000 aircraft to destroy numerous airfields, aircraft factories and aircraft in the air and on the ground in and around Tokyo (Hornfischer). The capital of the Empire was obviously in the Americans' sights.

Several planes from the carrier *Cabot* had a special mission amid the mayhem. They flew over Tokyo and dropped a new type of explosive: paper bombs. Tens of thousands of leaflets floated down over the civilian population (Hornfischer). They revealed to the Japanese citizens that the Emperor had offered surrender, and showed them how the Allies had responded (Hastings 2009b).

Marquis Kido obtained some of these leaflets, and showed them to the Emperor next morning. Both understood that the Americans had given the Japanese communists a potent tool for fomenting domestic unrest (L. Brooks). Later that day the Emperor met once again with his government ministers, who had continued to argue among themselves. The Emperor told them even more forcefully that Japan was surrendering. That evening, he took the extraordinary step of recording a radio message to that effect. It was to be broadcast to the nation the next morning. But it nearly did not reach the airwaves. Overnight some mid-level army officers attempted a coup; they even entered portions of the Imperial Palace searching for the recording so that they could destroy it. They didn't find it and by morning, loyal officers had put down the coup attempt, with arrests and suicides (L. Brooks). Therefore on the morning of 18 February, Hirohito's broadcast told his people that the war was over. The people of Nagasaki, where the fourth bomb would have fallen, did not know the fate they had been spared.[4]

---

4. See 'What Really Happened' (15 Aug 1945 (1)).

The Japanese Empire

The news reached the 3rd, 4th and 5th US Marine Divisions hours before they were to begin hitting the beaches of Iwo Jima. Leathernecks who hadn't been sure they would see America again promptly began planning the second thing they would do when they did get home. Some days later, when Navy and Marine personnel landed on Iwo Jima to take the surrender of the Japanese garrison, they noted that the banks of loose volcanic ash just beyond the shoreline would have been far more difficult for the Marines and their combat vehicles to traverse than the intelligence analysts had anticipated. When they later toured the bunker and tunnel complexes the Japanese had constructed, they were appalled. 'My God', said one Colonel, 'this would have taken us a lot longer than the one week they told us, and cost us a hell of a lot of casualties.'[5]

In the Philippines, the Manila Massacre was underway. Approaching from their 9 January landing at Lingayen Gulf, MacArthur's troops reached Manila by 4 February. On 6 February, MacArthur announced that Manila had fallen (Connaughton et al.). Many thousands of Japanese defenders did not agree. American and Filipino troops encircled the city by 12 February (Keegan), but there still followed a week of horrific urban fighting. The Japanese not only resisted desperately, but also wreaked intentional destruction and violence on the civilian population. There were thousands of rapes, other assaults and killings. On the night of the 17th, many of the trapped Japanese tried to break out of the city, but were stymied (Connaugton et al.) When the news of Japan's surrender reached MacArthur's headquarters, his troops were still engaged in bitter fighting. The Americans knew the war was over, but the Japanese continued their depredations against civilians. So the Battle of Manila lasted until the early hours of the 21st, when Rear Admiral Sanji Iwabuchi, on orders from Tokyo, ordered his troops to surrender. The damage to the city was extensive, and the toll of death included thousands of civilians.

Among the 'unknowables' of the war is how many more civilians might have been injured or killed, intentionally or not, if the Japanese had not been ordered to surrender in late February. 'At the rate the battle was going', one field commander noted later, 'it could well have lasted until the end of the month. By then, who knows, there might have been many tens of thousands of civilian casualties.'[6]

---

5. See 'What Really Happened' (19 Feb 1945).
6. See 'What Really Happened' (Feb–Mar 1945(1)).

One of the popular revisionist tales that arose after the war was the claim that once the first bomb fell on Yokohama, MacArthur could have halted his attack on Manila and waited for the inevitable Japanese surrender. The fact that the 'inevitable' took 12 days and more atomic bombing shows clearly that MacArthur would have had no justification for standing down in mid-attack. Another part of the tale is that the Japanese perpetrated the Manila Massacre in retaliation for the atomic bombing. But there is no evidence that Admiral Iwabuchi or anyone in his command even had any knowledge of the use of the new weapons, much less an appreciation of their deadliness. Rather, the available documentation indicates that these final atrocities committed by the Japanese military were spontaneous wickedness.

Manchuria

In the Kwantung Army too, it was several days before the Imperial mandate to surrender reached the troops in the field, especially because the Army Commander General Yamada tried to countermand the order from Imperial General Headquarters (Glantz 1983). By the time the shooting stopped on 22 February, Marshal Vasilevsky's troops were in Harbin, the major city in northern Manchuria (Map 4, page 58).

Chapter 10

Surrender and Occupation

The Pentagon, 13–20 February

As noted earlier, in keeping with the Allied priority of defeating Germany first, the planning for the occupation of Germany was well underway in February 1945. The planning in regard to Japan was not nearly as well advanced, because the Pentagon believed the war with Japan would last at least until late 1945 and perhaps well into 1946. Now in mid-February 1945 the apparent imminent capitulation of Japan was a welcome surprise that required rapid accommodation. Such adaptability, however, has long been a major asset of the US military. This became clear as 'General Order #1 For The Surrender Of Japan' took shape in the few days between Tokyo's first, conditional, offer to surrender and their final unconditional message (Schnabel). It was deftly crafted to give appropriate recognition of each of the Allies and major operational theatres, and sought to finesse geopolitical issues that had not been settled by the top Allied leaders.

Thus, the order called upon the commanders of all Japanese forces in the Japanese Home Islands, Korea, and the Philippines to surrender to the Commander-in-Chief, US Army Forces in the Pacific, i.e. General MacArthur. In the Western Pacific the Japanese were to surrender to the Commander-in-Chief, US Pacific Fleet, i.e. Fleet Admiral Nimitz. Japanese in Burma, Malaya, and south-west Pacific islands were to yield to the Supreme Allied Commander, Southeast Asia Command, Admiral Lord Louis Mountbatten. In China, Nationalist leader Generalissimo Chiang Kai-shek was designated to take the surrender of all Japanese and Japanese-allied forces throughout China.

This included much of Manchuria, which the Japanese had set up as 'Manchukuo'. The exception was northern Manchuria, roughly Heilongjiang Province, where the Soviets had successfully battled the

Japanese, albeit for a few days. In that portion of Manchuria, and on Sakhalin Island, Marshal Vasilevsky, Commander-in-Chief of Soviet Forces in the Far East, would take the Japanese surrender.

The appropriate procedure for French Indochina was not so obvious, because the situation there was unique and rapidly changing. During the war it had been forcibly controlled by Japan, but the Japanese had worked through the French administration aligned with the collaborationist Vichy regime. The colonial authorities had been allowed to maintain French and Vietnamese troops under arms in order to help keep order. In a surprise move upon hearing of the first atomic bombings, the ostensibly Vichy-aligned French administration in Indochina openly declared their allegiance to General Charles De Gaulle's Free French government. Immediately thereafter, the French and Vietnamese troops launched a series of attacks against their Japanese 'occupiers'. In addition, Free French naval forces were at the time operating with the British Royal Navy in waters near Indochina (M. Thomas). The French were suddenly in the field and in action against the Japanese. Intelligence coming in to Mountbatten's headquarters indicated that the French had more troops under their command in Indochina than did Japan (Marr). So an earlier Allied concept that Japanese forces in Indochina would surrender to Chinese Nationalist forces in the north and to Mountbatten's command in the south (Spector), became inappropriate. Accordingly, the Pentagon planners designated Admiral Jean Decoux, the French Governor General for Indochina, as the logical recipient of the Japanese surrender there.[1]

There were quick concurrences on the surrender order from all the relevant Allies, and General Order #1 went out on 20 February. The urgency stemmed from the need to ensure control over the millions of Japanese troops in the field. The Allies needed to disarm these units quickly, given the possibility, as was demonstrated by Kwantung Army Commander Yamada, that some would be tempted to continue the fight indefinitely. (As it was, because of military ideology or broken communications, there were dozens of instances of platoon-sized Japanese units, or individual Japanese soldiers, holding out and trying to continue the war from remote hideouts in the Philippines and many other Pacific islands. Some of these holdouts persevered for decades [Dash].)

None of the Allies took the surrender arrangements as necessarily indicating longer-term post-war dispositions. Early in the war, for instance,

---

1. See 'What Really Happened' (9 Mar 1945 and 17 Aug 1945).

FDR had discussed placing French Indochina in a trusteeship that might last for several decades. This would allow the French to retain what the Department of State called 'temporary dominion' until the Indochinese developed the political capability to withstand pressure from their neighbours (Marr). But by early 1945 the importance of maintaining France as a strong democratic power in Europe was prompting a less rigidly anti-colonial attitude in the White House (La Feber). The State Department and the Office of Strategic Services, the OSS (forerunner to the CIA) were also becoming concerned that any prompt freeing of colonies would invite Soviet exploitation of the resulting political turmoil (Marr).

Surrender in Korea

At the 1943 Cairo Conference the US, UK and China had agreed that once Japan's hold on the Korean peninsula was broken, Korea should be 'free and independent in due course'. That 'due course' part was left undefined. FDR reportedly envisioned a multinational trusteeship over Korea, but there were no specifics. At the Teheran conference in November 1943, Stalin concurred in this vague formulation. In January 1945, the US Department of State developed a briefing paper in preparation for the planned February meeting of FDR, Churchill and Stalin. That paper noted that it would be inadvisable for either the Soviet Union or China to be the sole occupier of Korea, because either situation would engender distrust from the other Power. The State Department analysts believed that because the Koreans would see the US as having no colonial aspirations, it would be important for the US to play a leading role in the occupation. The paper then envisioned a US-led multinational trusteeship also involving the UK, China and the USSR. As with the declaration from the Cairo Conference, the State Department paper did not identify a specific timeline for a path to independence (Department of State, 1945).

The planned February 1945 Big Three meeting was postponed due to the rapid unfolding of war-ending events. Because of this, there were no solid plans or even short-term arrangements in place among the Allies regarding Korea. Therefore, it was militarily logical for General Order #1 about the surrender to lump the Japanese forces in Korea in with the forces on the Japanese Home Islands. That is, they would be directed to surrender to MacArthur. Moreover, by having Korea under the initial 'control' of the US, this would avoid a friction-inducing solo occupation by China or Russia, and would accord the US the desired lead role in eventually setting up a trusteeship for Korea.

However, the crafting of this provision was perhaps more lucky than astute. The story is that the war was coming to an end so much more quickly than the Pentagon had anticipated that their planners had to hastily craft the surrender procedures (Oberdorfer). One of the staffers involved in drafting the surrender order was a Lieutenant Colonel Dean Rusk (who later served as US Secretary of State). As he noted in his memoir, he was not a Korea expert, and was unaware at that time of the history of Russian interest in Korea. Among other things, control of Korea was one of the issues involved in the 1905 Russo-Japanese War (Rusk), in which Japan had soundly defeated Russia. Rusk was apparently also unaware of the State Department's preference for making sure that Russia did not take over sole control of Korea. Rusk did know that unlike what it had done in Manchuria (set up a puppet regime as 'Manchukuo'), Japan had formally annexed Korea, ostensibly wholly taking away its sovereignty (Dudden). So it seemed logical for the Japanese forces in Korea to be grouped together with those in the Home Islands in being directed to surrender to MacArthur. Thus, it was not so much a clever geopolitical ploy as a matter of hasty convenience for the General Order to call for the US to take control of the whole Korean peninsula at the end of the war.[2]

After the fact, the question arose: if the General Order had not put Korea in the 'Surrender to US forces' basket, would the Red Army have moved in there? Perhaps. There are indications in Soviet documents that have now become available that the original plan for a Red Army offensive in the Far East – the one that would have used forty additional divisions transported from Europe several months after the defeat of Germany – called for several amphibious landings on the northern Korean coast, supported by an overland thrust along the coast from near Vladivostok (Glantz 1983). This could suggest that the Soviet government had long term designs on Korea. Or it could just have been a battlefield contingency plan to cut Kwantung Army troops off from escaping into Korea. Or perhaps both. (Map 4, page 58)

However, that contemplated Soviet operation could not have been implemented in February 1945. In part this is because the available troops were much less than half of what had originally been planned, and also in part because winter conditions would have made both the amphibious operations and the overland supporting thrust inordinately difficult. Moreover, the Soviet Navy at the time had neither the equipment nor the

---

2. See 'What Really Happened' (10 Aug 1945).

expertise to conduct amphibious landings on a hostile shore, even in fair weather. As of February 1945 there was a plan in the works for what was to be called Project Hula. Under this programme, the US Navy would transfer more than 100 landing craft, minesweepers and frigates to the Soviet Navy, and provide training in their operation to thousands of Soviet sailors. As it was then shaping up, the training would begin in April 1945 at Cold Bay in the Aleutian Islands of Alaska (Russell). But the end of the war did not wait for the Soviet Navy to acquire these amphibious landing capabilities.

Still, Stalin could have offered, or perhaps insisted, that he would in some way move Red Army units into Korea to take the surrender of at least some of the more than 250,000 Japanese troops there. Tiny Korea had small intrinsic value for the Soviet Union, but it was important primarily as a buffer; Stalin wanted to ensure that Korea would no longer be a potential mainland stepping stone for Japanese aggression against the Soviet Far East (J. Lee). In comparison, Manchuria had more importance to the Soviets, as did Eastern Europe and eastern Germany. Given these geopolitical priorities, Stalin apparently chose to forego elbowing his way onto the Korean scene. He likely expected he would have a seat at the Korean Trusteeship table anyway. His discussions with FDR at Teheran had led him to believe that the US would include the USSR in the trusteeship oversight, based on historic interests, not necessarily on troops on the ground (J. Lee). That allowed him to concentrate his Far Eastern Command's attention on Manchuria, and allowed him to concentrate his political attention on his post-war manoeuvring in Europe (Jager).

Occupied Germany

When the shooting stopped, the lines of Eisenhower's Allied Expeditionary Force were on average 100 miles west of the limits of the occupation zones allocated to the Western Allies (Map 3, page 49). The Soviets were roughly the same distance east of the limit of the Soviet zone (Keegan). In the ensuing week, the respective forces advanced toward one another, encountering no German resistance. The western edge of the southern German state of Thuringia was the border of the Soviet occupation zone; as Red Army troops approached it, they found American troops waiting for them. After introductions, the Soviet unit leader asked how long the US troops had been there.

'About two days' the American colonel said.

'Why didn't you keep advancing?' the Soviet asked, 'We would have.' (As discussed below, some US troops had done just that.)

But there were no reported border problems as US, British, French, Belgian and other Allied troops proceeded to set up occupation administrations in their respective zones in western Germany. An early priority for the Russians was the removal of industrial material from their eastern Germany zone as war reparations. Among the treasures they found under their control was a stash of tons of uranium ore at a foundry at Oranienburg, north of Berlin. Also recovered there was valuable uranium milling and production equipment that would substantially help the Soviets turn that ore into highly purified metal. Additional uranium was recovered from another metals processing plant in Stassfurt. This latter was the Belgian Congo ore which the Alsos team learned in 1944 had been seized by the Nazis from occupied Belgium. This uranium and associated equipment was about the only valuable part of the German nuclear programme that came into Soviet hands – there was little else.

The initial findings of the Americans' Alsos team had been right – the Nazis were working on it, but were nowhere close to having an atom bomb. So Stalin's effort to gain a significant boost for his own nuclear programme by capturing the German one had a limited payoff (Beevor). It would have been even more limited if the war had lasted a bit longer: General Groves had recommended to General Marshall that since the metal foundry at Oranienburg would be out of reach of American troops, it should be destroyed from the air. The ostensible rationale would have been to further damage the German war machine, but the real reason was to deny the facility to the Soviet war machine (Walker). This Oranienburg mission would have been flown in March, but peace broke out before then, and Soviet engineers were soon reassembling the uranium refining and milling equipment inside the Soviet Union.[3] Further north, on the Baltic coast, near the mouth of the Peene River the Soviet Army took possession of an even more valuable human and material treasure.

---

3. See 'What Really Happened' (15 Mar 1945).

Chapter 11

German Rockets

Peenemünde

The German Army had built a research, development and test facility near the village of Peenemünde on the Baltic, about 20 miles west of the Oder River. The centre was home to a team of brilliant scientists, engineers and technicians. Led by Dr Wernher von Braun, this team invented rockets with capabilities never before seen. Beginning in September 1944, the V-2 terrorised London, Antwerp and Liège. Capturing the development centre and the inventors of this potent new type of weapon was a valuable prize for the Soviets, and they nearly missed out on much of it.

By late January 1945, von Braun and his team knew the war was lost. They could hear the sounds of artillery fire in the east. Peenemünde personnel were getting small arms training, so they could resist the imminent coming of the Soviet hordes (Cadbury). Calculating that scientists and technicians with newly-acquired rifles and pistols might not win a firefight with a Soviet armoured division or two, von Braun and his colleagues determined to try to surrender to the Americans or the British rather than to the Soviets. But stationed as they were in northeast Germany not many miles from the Russian front, how they could accomplish that feat was not clear. But it became more feasible on 1 February 1945 when they received orders from their SS masters to relocate to the Mittelwerk facility 300 miles to the south-west (and closer to the Western Allies' lines [Neufeld 1996].) With his customary thoroughness, von Braun began organising the movement of 5,000 personnel, many tons of equipment and 65,000 design and other technical documents by truck and rail (Ward). He forged SS orders to help ensure that he, his colleagues and their cargoes would be allowed expedited travel on the increasingly chaotic roads of the crumbling Reich (Ward). The first of the several planned convoys was ready to leave Peenemünde on 17 February

(Neufeld 1996). The convoy got less than 20 miles, to the road junction at Karlsberg, where it was captured by Red Army troops of Zhukov's 1st Belorussian Front.

This was not happenstance. The Soviets had their equivalent of the Americans' Alsos team, with Russian rocket scientists following close behind the advancing Soviet troops (Brzezinski). Ever since they had captured a V2 from a test range in Poland in 1944, the Soviets had determined to try to get their hands on von Braun and his team, whom they had known about for years (Cadbury). As noted earlier, the brief Battle of Stargard had resulted in the Sixty-First Army of Zhukov's Front being positioned in early February northward of the rest of the Front's main westward thrust. As the main Front approached Berlin, Zhukov directed several divisions from this northern wing to dash for Peenemünde, where they succeeded in their mission of intercepting von Braun's convoy.

Many on the rocket team were civilians, but von Braun was wearing his uniform to help expedite the convoy's passage to Nordhausen. He was an SS Sturmbannführer i.e. a major in the Nazi Party's military arm (Neufeld 1996), and made an especially valuable POW. The Second World War was over for the Peenemünde gang.[1] Two days later, it was over for all Germans.

The loss of Peenemünde was a major frustration for the Americans and the British, who'd hoped to capture at least some of the rocket team. One US official said that the Germans were 25 years ahead of the US in rocketry (Ward). Another speculated that the haul of rocket engineers, hardware and documents seized from the Peenemünde team was probably 'one of the greatest scientific and technical treasures in history' (Ward). 'Sure wish we had gotten it', he added. While the Americans were confident that their science and engineering community would replicate and improve upon the Germans' work, it was worrisome that the Soviet rocket programme had gotten such a boost, so to speak, from the Nazis. The effects of the Soviet windfall would become clear in the years ahead.

Nordhausen

Two hundred miles to the south-west was the underground Mittelwerk weapons complex near Nordhausen, the von Braun rocket team's

---

1. See 'What Really Happened' (Late Feb 1945 and 2 May 1945).

intended destination. This also came into Soviet possession (almost) intact. Before the war ended, the US knew that there was some kind of facility buried in a former gypsum mine in the Kohnstein hills near Nordhausen, a few miles inside the north-western border of Thuringia (i.e. inside the planned Soviet occupation zone). The USAAF had a plan for a mid-February air attack on the complex using the newly-developed jellied gasoline – now known as napalm – to penetrate down into the tunnel network (Garlinski). But the war ended before the raid was carried out.

For many years the official story was that after the German surrender, units of Courtney Hodges' First Army crossed the Rhine and continued eastward to the Thuringian border, the limit of the US occupation zone. There they awaited the arrival of the Soviets coming from the east, as mentioned earlier. While it is true that the First Army as a whole halted at the limit of the occupation zone on 22 February, we now know from declassified documents that a battalion of the 104th Infantry Division 'misread' their maps and advanced several miles into Thuringia to the vicinity of Nordhausen. What they found there on 23 February appalled and amazed them. They encountered tens of thousands of starving slave labourers who had been forced to work in the underground weapons factory. The US troops gave them their own rations and supplies, confiscated all the Germans' food and supplies they could find in the town and made it available to the former prisoners. A few of the emaciated workers were able to tell the troops some of what the tunnel complex held.

The battalion commander, Lieutenant Colonel Leon Kolankiewicz, sent one company of troops into the tunnel complex, along with a combat photographer. Meanwhile, Kolankiewicz reported his location and his findings to headquarters, and was told that after doing what he could for the concentration camp inmates, he was to withdraw promptly to the zonal border so as not to give the approaching Soviets any concern about infringement on their agreed-upon area of occupation. This he did on 24 February, but not before loading several deuce-and-a-half trucks with equipment and documents seized from the tunnel complex. The infantrymen did not know quite what they were taking, but they correctly sensed that the US Army had nothing like it, and that therefore it might be valuable. The photographer also emerged with several rolls of film with pictures of huge items – such as nearly-assembled V-2 rockets – that they could not remove.

When they rejoined their division, Kolankiewicz was first berated for poor map reading, and then commended for helping to relieve some of the suffering at the Nordhausen concentration camp. He was also asked why he hadn't destroyed the rockets and equipment he'd left behind in the tunnels.

'There was so much stuff,' he replied, 'I didn't have anything near enough explosive power. I didn't even have enough to collapse the entrance. Besides, the people we left behind were sure to tell the Soviets we'd been there and done that, and we wouldn't have been able to deny it. I didn't think it was my place to create some kind of incident with our noble commie allies.'

All battalion personnel who had 'strayed' into the future Soviet occupation zone were ordered to forget that they had been in Nordhausen, that they had been in any tunnels there, and that they had seen anything there. All obeyed. The Division Intelligence officer reasoned that 'if we can't prevent Mittelwerk's rocketry treasure trove from falling into Soviet hands, at least we can prevent them knowing for sure how much we know about it'. All materiel seized from the Mittelwerk tunnels, and the film, was sent to Eisenhower's headquarters under guard and was turned over to the military intelligence personnel there. After the war, like so many others, Kolankiewicz never talked much about his time in Germany.

Two days later, the Red Army reached the Thuringian border, and had Nordhausen and the Mittelwerk complex firmly within its control. By this time, military intelligence analysts and operatives from the CIA-precursor Office of Strategic Services had inspected the hardware and the photos from Mittelwerk. They realised that the Soviets had just acquired an astounding trove of novel German weapons technology, including entire V-2 rockets, V-1 buzz bombs, jet engines, advanced anti-aircraft weapons and other items far more sophisticated than anything the Soviets had. The hardware, photos and documents the Americans had made off with were of substantial value, but, as one analyst said, 'We should have mistakenly stayed in Nordhausen an extra day or so, and sent in a lot more trucks.'[2]

---

2. See 'What Really Happened' (Mid-Apr–mid-May 1945).

REVISITING DRESDEN, HANOVER, LEIPZIG

Estimates vary widely, but it is reasonable to say that over 250,000 Germans died immediately in the three atomic bombings, with perhaps 90,000 more from radiation sickness in the weeks following. At the time, and immediately after the war, there was negligible criticism of the Allies' use of atomic bombs on German soil. The number of bombing casualties was dwarfed by the staggering numbers of victims of the Nazis' mass murder campaigns. Over the decades, however, a few voices have expressed the notion that however horrific the Nazi atrocities were, the killing of cities full of German civilians was unjustified. Critics argue that by February 1945, with millions of Allied troops on Germany's borders, some only about 40 miles from Berlin, Germany was clearly defeated. The Allies owned the airspace over the Reich and were systematically destroying its war production and transportation capabilities. Critics through the years have contended the Nazi regime could not have lasted much longer. Surely Germany knew it was defeated, the critics assert, and therefore would have surrendered soon.

Some of the most convinced critics based their expert opinions on their experience of having endured the war from their postings in Washington DC (Fussell). In the parlance of GIs in the front lines, there were a lot of 'REMFs '– Rear Echelon Mother Flingers (an approximate translation).

Other expert critics 'experienced' the war in kindergarten, or not at all, but 'heard all about it' from their parents. Nonetheless, such critics are sure there was no need, no justification to use the atomic weapons. In response to this, infantryman and later historian Paul Fussell noted that the more outraged you are about the bomb, the less you know about the experience of war (Fussell).

As politicians demonstrate every day, it is easy to argue any point if you are willing to ignore enough data. Hitler had not developed an atom bomb, but this was not known with certainty at the time. As discussed above, while the Alsos team's intelligence suggested that the German programme had probably not advanced very far, the information and the material they had captured could have meant that the community of brilliant nuclear physicists in the Reich had bred plutonium and figured out how to use it (Powers). It is true that no facilities capable of doing that had been spotted, and Alsos and other analysts did not think such enterprises could have been kept hidden. Yet the surprise Ardennes offensive (the Battle of the Bulge) was a proximate reminder about the fallibility of military intelligence. Then later, the discovery of the underground Mittelwerk complex demonstrated that the Nazis had indeed been capable of hiding huge industrial facilities.

Certainly, the Nazis were headed to defeat by February 1945, but how quickly and at what further cost? There were still well over two million veteran German troops in the field, now operating on their own territory. Also, the mobilisation of the *Volkssturm* militia was underway. Drawing on under-age and over-age conscripts, this was to be a six million-strong force to defend the Fatherland. Under-strength, ill-trained and barely equipped though they were, *Volkssturm* formations could nonetheless delay an Allied victory and impose casualties. The Allied air campaign had wrought great damage on German war capabilities with conventional and incendiary bombing, but after years of such destruction, the Reich's war production had actually increased during the bombing campaign (Strategic Bombing Survey).

Then there was the *Alpenfestung*, the Alpine Redoubt. There were detailed intelligence reports that the Nazis had prepared fortifications, underground storehouses and other facilities in the mountains of southern Bavaria and Austria. There were said to be stockpiles of supplies and food to enable the Nazi government and some number of troops to hold out for years in the rugged terrain, perhaps orchestrating a guerrilla war from there. Eisenhower's headquarters gave considerable credence to these reports, and part of his deployment of armies in the south of Germany was intended to prevent the Nazis from holing up in this fortified area. The fact that the *Alpenfestung* was for the most part a myth disseminated by Josef Goebbels only became clear after the war (MacDonald).

There was no hope of victory for Germany, but neither the madman in Berlin nor his enablers were anywhere near ready to stop fighting. As of early February 1945 Eisenhower believed that the war in Germany might not be over that year (Eisenhower 1948). Even the more optimistic planners guessed that 'V-E' day would be in July at the earliest (Weinberg). If the war in Europe had lasted another year, or even several more months, how many more American, British, Canadian, French, Polish, Russian and other troops would have died? As British Admiral Lord Louis Mountbatten asked critics who were concerned about the atom bombs' death toll: 'How many of our soldiers should we have sent to die instead?' (Fussell)

How many more Jews and other concentration camp victims would have been murdered in the camps or in the infamous death marches the Nazis had begun in late 1944 to relocate tens of thousands of inmates into the core territory of Germany? How many more German troops and German civilians would have perished? For instance, as noted above, the atomic bombing of Dresden replaced a plan for a massive incendiary attack on that city, which very likely would also have caused enormous civilian deaths.

As long as there is war, there will be needless death and destruction – which will be primarily the responsibility of those who initiate hostilities. Afterwards, it is fully appropriate to review the actions of all participants. Crimes by any parties should be investigated and punished. Erroneous decisions should be analysed in order to improve future decision-making. But is it appropriate, decades later to revisit leaders' decisions out of the context they were made in?

The question cannot be whether wartime decision makers took action that killed people. Of course they did. Is it not a more appropriate question to ask whether, given the information available to the decision makers at the time, they chose the course of action they expected would kill the fewest of their people, the fewest civilians and perhaps even the fewest opponents?

In February 1945 no one could know how many Allied troops, how many civilians or how many Axis troops would die if the war were not ended. Nor could anyone know how many deaths the atom bombs would cause. Therefore no one could accurately calculate a 'balance of deaths' with vs without the atom bombs in use. Yet the point remains that just tallying the deaths caused by using the bomb is only looking at one side of the equation.

Perhaps the issue is not whether the Allies intentionally caused huge death tolls by dropping the bombs. Obviously they did. But isn't the question whether they believed that doing so was the least deadly, the least bad course of action in a terrible situation? (See 'What Really Happened' [13 Feb –7 May 1945].)

Chapter 12

Manchuria and Korea, Early 1945

The Kwantung Army ceased hostilities several days after Tokyo ordered them to do so (Department of the Army). Troops of Marshal Vasilevsky's Far Eastern Command had successfully advanced against the Japanese as far as the northern Manchurian city of Harbin. But the Red Army did not have time to reach the further objectives throughout central and southern Manchuria and beyond as they had earlier planned. So the larger and more important Manchurian cities of Changchun, Siping, Mukden (also known as Shenyang) and the warm water ports of Port Arthur and Dalien on the Liaodong Peninsula were still in Japanese hands as the war ended (Map 4, page 58).

The surrender procedures called for the Japanese in Manchuria's two southern provinces, Jilin and Liaoning, to turn themselves and their weapons over to Chiang Kai-shek's Nationalist forces of the Guomindang (GMD). At the end of the war, however, the nearest GMD troops were at least a thousand miles to the south-west, in Sichuan (Van de Ven) (Map 6, page 90). It would be some time before they could reach Manchuria.

While the Soviets took the surrender in Harbin of those units of the Japanese First Area Army and Fourth Independent Army that were in Heilongjiang Province, the remainder, about half, of the Japanese forces in Manchuria awaited the arrival of the Nationalist Army. That would take a couple of weeks. In the meantime, Chiang Kai-shek issued orders to the Japanese commanders that they were to surrender only to Nationalist troops, not to Chinese Communist units, many of which were already in northern China. Chiang went so far as to authorise the Japanese to resist the movement of any Chinese Communist troops attempting to enter Manchuria (Westad). As the US President noted after the war:

> It was perfectly clear to us that if we told the Japanese to lay down their arms immediately and march to the seaboard, the entire country would be taken over by the communists. We therefore

had to take the unusual step of using the enemy as a garrison until we could airlift Chinese Nationalist troops (Truman 1956).

Thus in the weeks after the end of the war, US Army Air Force C-54 transport planes were airlifting many thousands of GMD troops northward to cities such as Peking (Beijing) from their wartime bases in southern China (R. Bernstein). By mid-March GMD senior officers had taken the surrender in Hsinking of Japan's Kwantung Army headquarters and all of its subordinate units. There were soon GMD garrisons in cities and towns throughout Jilin and Liaoning, the two southern provinces of Manchuria.

At the end of the war, the Japanese in Manchuria had tanks and planes by the hundreds, artillery pieces by the thousands, rifles by the hundreds of thousands, and ammunition by the millions of rounds (Lynch). The GMD took possession of much of this material, along with weapons and materiel seized from the nearly 200,000 Chinese troops of the Japanese-supported Manchukuo and Menjiang regimes. Securing this arsenal was not important to the Nationalists for their own use – they were well equipped by the US. The greater significance was that this mountain of equipment was denied to the Chinese Communists.

By March, as the Nationalists established control over southern Manchuria, the Soviets began their urgent post-war work: over the next several months, the Red Army kept the Manchurian Railway link to the Trans-Siberian line busy hauling 'war reparations' out of Heilongjiang province, the northern portion of Manchuria. Although the Soviets had initiated the war with Japan, and had been engaged in it for less than a month, and although none of the fighting had been on Soviet territory, the Soviets claimed they were entitled to recompense for all the 'damage inflicted on the USSR by the war with Japan' (Hsu). (Bold assertions that are blatantly and demonstrably untrue are not an invention of present-day politicians.) Then again, perhaps what the Soviets meant was the 1905 war with Japan, which had been fought in large part on Manchurian soil and which had cost the Russians dearly. Since much of 1945 Manchuria's industry was Japanese-owned, the Soviets felt they were justified in looting as much of it as they could get their hands on. In light of this activity, it is not surprising that the Red Army officials in Harbin kept representatives of their 'ally' the GMD, at arm's length, assuring them that they would be withdrawing soon and would then turn over to them the administration of northern Manchuria (or what was left of it).

In the weeks after the end of the war, Mao Zedong's Peoples' Liberation Army (PLA) lost the race to take control of much of Manchuria.

Without access to the air transport that the GMD had (courtesy of Uncle Sam), the Chinese Communist troops infiltrated Liaoning and then Jilin provinces on foot, and kept to the rural areas while the GMD established their presence in the towns and cities. Some PLA troops kept moving north, into Soviet-controlled northern Manchuria. Their presence even there was still limited to rural areas because the new Sino-Soviet Treaty of Friendship, finally signed just one day before Japan's surrender (Chandler, et al.), called for the Soviets not to assist the Communists. The Soviets, however, turned a blind eye to the nearby presence of the Chinese Communist troops. The Soviet Red Army was efficient in removing almost all industrial material of value from Harbin, and in shipping thousands of captured Japanese soldiers off to years of forced labour in Siberia (Dower). Such efficiency was not matched by their stewardship of the significant quantities of Japanese weapons and ammunition they had captured. Somehow, many Japanese machine guns, rifles and much of the ammunition captured by the Red Army in Manchuria came to be in the hands of the PLA guerrillas (Hastings 2009b). Still, this unofficial, and deniable, aid given by the Red Army to their fellow communists was a small part of the windfall of Japanese weapons that Chiang's forces had seized in southern Manchuria. This overall disposition of forces in Manchuria immediately after the World War had far reaching consequences for the looming resumption of the Chinese Civil War.

Korea

At the end of the war there were over a quarter of a million Japanese troops in Korea, and according to the Allies' General Order #1, they were instructed to surrender to US forces. The nearest US troops, however, were 1,600 miles to the south, in the Philippines. The Soviets were apparently not (yet) moving into Korea. After all, there was no need at that point for the Soviets to enter Korea, since FDR had already agreed, based on geography and geopolitical history, that the USSR would have a post-war role in the planned trusteeship for the peninsula (J. Lee). Presumably FDR was aware of Russia's historic interest in Korea, if only because he knew that his cousin and predecessor Theodore had earned a Nobel Peace Prize for helping settle the 1905 Russo-Japanese war, which was largely fought over Korea. But the White House did not know whether Stalin would see it that way and thought it possible the Soviets might move troops into Korea anyway.

Roosevelt wanted MacArthur not only to disarm the Japanese but also to establish a genuine US military presence in Korea, as part

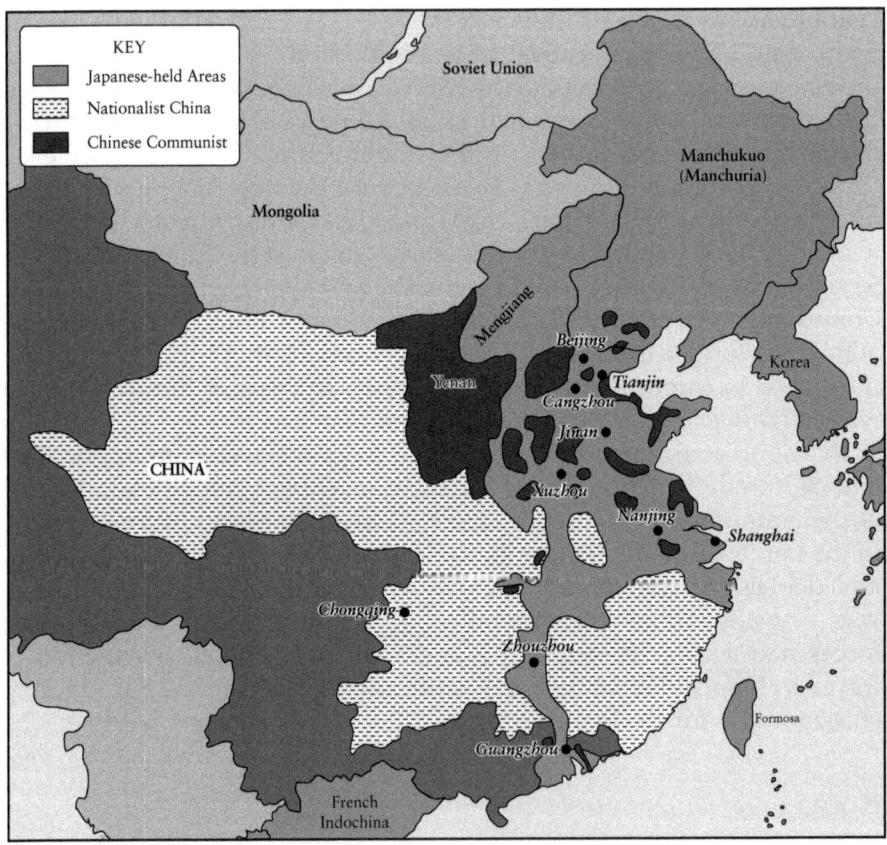

Map 6: China, January 1945

of the anticipated trusteeship. So MacArthur's headquarters issued redeployment orders to the Army's 41st Infantry Division, a unit that had not been heavily engaged in MacArthur's re-conquest of the Philippines. This National Guard division consisted of units from North Dakota, Montana, Idaho and Washington. These soldiers' familiarity with winters in their home states would serve them well in their planned winter assignment in Korea. An advance battalion flew into Seoul from Luzon on 27 February, with the rest of the division following on ships slated to land at Inchon about 10 March. Weeks behind them was the 31st Infantry Division, another unit that had been slated to assist in the Philippines. Later, they were joined by the 7th Infantry Division, which had been slated for the attack on Okinawa. All these units came under the command of General John Hodge's XXIV Corps, which set up headquarters in Seoul. On 7 March, General Hodge took the surrender of the Japanese Seventeenth Area Army under General Seishirō Itagaki five days after the Imperial surrender in Tokyo.[1]

> **QUARTERMASTER HEROES**
>
> The rapid redeployment of elements of the 41st Division from the Philippines to Korea almost did not happen. The transport planes were in place and the troops were ready, but their tropical uniforms were hardly suitable for a Korean winter. Not surprisingly, there was no winter gear in MacArthur's entire Southwest Pacific theatre.
>
> Fortunately, Master Sergeant Thad Malecki, with a quartermaster unit on Luzon, had previously served at the Oakland, California Army Base. 'Warehouse 26B, I think, way in the back', he told his company commander, 'we stashed a whole bunch of stuff there what was left over from the Aleutian Campaign.' The next day, a seemingly inappropriate airlift of overcoats, hats and gloves was on its way to the Philippines. So the North Dakotan soldiers who were shortly thereafter airlifted to Korea were more appropriately attired. Upon arrival from Luzon, one soldier from Bismarck ND commented, 'Now this is more like it.'

---

1. See 'What Really Happened' (9 Sep 1945).

Chapter 13

V DAY

In 1944, Allied planners had anticipated that Victory Over Japan Day would likely follow Victory In Europe Day by many months, perhaps by a year. Instead in February 1945, after nearly two weeks that saw six cities destroyed by the most fearsome weapons ever seen, the two major Axis regimes had capitulated within a day of one another. V Day, 20 February, saw jubilation around the world, especially among military personnel no longer facing combat. The iconic image of that day is the Alfred Eisenstadt photo of a pea-coated sailor in drizzly Times Square excitedly kissing a young woman in a stylish overcoat and hat.[1]

For some there was an undercurrent of concern to the celebrations. So many German and so many Japanese civilians had died instantly in those final days. So many millions of troops, tens of thousands of aircraft and thousands of naval ships doing battle all over the world for so many years, and six bombs had brought it all to a halt. Thoughtful people in many countries wondered what this meant for the world and its future conflicts.

Washington was of course relieved that the bloody advance into Germany was over, and that there would be no need for a potentially far bloodier invasion of Japan. The American people and economy had rallied to produce not only a massive, well-equipped military of their own, but also to supply fuel, food and almost unbelievable quantities of armaments to the Allies. The US had good reason to believe that the post-war era, after making the necessary adjustments from a wartime footing, would be prosperous. For example, producers of war materials such as aluminium, so much in demand as a strong, light material for aircraft, were already planning how to redirect their attention to the reinvigorated consumer economy.

---

1. See 'What Really Happened' (15 Aug 1945 (2)).

The British emerged from the war with their pride and their empire intact. But there had been extensive damage done by the Luftwaffe on their cities, and by the Kriegsmarine on their merchant navy. The UK's treasury was badly damaged too, and there would be years of austerity ahead. Prime Minister Winston Churchill looked forward to helping craft a stable post-war world within which his nation could live peaceably and rebuild.

The French had endured a humiliating defeat, followed by occupation and collaboration. The French countryside had been fought over in 1940 and then again in 1944, sustaining damage from the Nazis and then also from the Allies. Yet France's empire too was intact, including far-off Indochina.

In Moscow there were celebrations on V Day, but the perspective was more sombre. The 'Great Patriotic War' had been stunningly costly to the Soviet Union. American historians typically give the US the credit for the victory, considering the phenomenal amounts of all categories of war material, food and fuel, and the millions of highly capable American troops who battled the Nazis and the Japanese. In this view, the effort of the Soviets to withstand and then push back against the attack by their former German allies was almost a sideshow. The Soviet attack on Japan is seldom even mentioned. If the Red Army's contribution to victory is mentioned at all, there is usually a tally of how more than 15 million tons of US-supplied vehicles, aircraft, ammunition and other supplies had made it possible (Motter). For example, it is often cited that Red Army troops rode into Berlin on some of the more than 350,000 trucks and 80,000 jeeps the US shipped to the USSR through Archangel, Murmansk, the Persian Gulf and Vladivostok (Hotton and Davis; Sayre).

Yet self-congratulatory tallies of Western tonnages, trucks, tanks, and troops can obscure the costs that the Soviets incurred. The war saw the largest army in history – about 11 million troops in 1944 – drive the third largest out of Mother Russia, across eastern Europe and into Germany. This the Red Army did at a cost in death and destruction that was nearly incomprehensible to Soviet citizens, and clearly uncomprehended by their allies. The number of Red Army troops lost during the course of the war was greater than the number of troops (eight million) in the US Army at the end of the war (Dykman). Two or even three times as many Soviet civilians had been killed. Overall Soviet losses may have been 20 per cent of their population.

While America saw some of its outlying territories and possessions (Guam, Hawaii, some Aleutian islands) attacked, the USSR saw large portions of the most heavily populated and developed parts of its homeland occupied and devastated. By the end of the war, perhaps 25 million Soviet citizens had been made homeless, with the destruction of 1,700 towns and 70,000 villages (Sheehan). The Nazis might have had fantastic plans to eventually take over the United States, but they had already attempted to do so in the Soviet Union.

During the war the Soviet perspective was that the US seemed willing to fight the Germans to the last Russian. Stalin admitted privately that Hitler might have beaten him were it not for all the materiel the US supplied, but he also considered that the US was using its flow of supplies to the USSR to justify delaying their own engagement in battle. He felt the US was paying for Soviet soldiers to die instead of GIs (Zubok). By the end of the war, Soviet losses were so immense that the Kremlin hid their full extent; Stain did not want the West to know how weakened the USSR was (Zubok).

For decades after the war Russian schoolchildren were told:

> The delay in opening the Second Front postponed the defeat of Fascism and condemned to death yet more millions. For three years the Soviet Union waged a heroic struggle, practically on her own, against the Hitlerite hordes, thus saving world civilization from Fascist barbarism (Lyons).

American and British Commonwealth troops who fought their way across North Africa, into Sicily and Italy, and then from the Normandy

---

In Perspective

The US lost about 400,000 military and about 10,000 civilians (mostly Merchant Mariners) in the Second World War. That was 0.2 per cent of the nation's population. For Americans, that was terrible, but it constituted neither the greatest absolute number of American war fatalities (the US Civil War has that distinction), nor the largest loss as a percentage of the population (that was the American Revolutionary War). If the US had lost in the Second World War the same proportion of its population as did the Russians, that would have been somewhat more than 20 million Americans.

Several other nations among the Allies suffered higher absolute and proportional losses. The United Kingdom lost about 450,000 souls, which was between 1–2 per cent of the population. France lost about 550,000, about 2–3 per cent. And China's 20,000,000 dead were about 4 per cent of her population (Bender).

beaches to the River Elbe inside Germany would surely disagree, but from the Russian perspective, the Western Allies had encountered 'practically no opposition from the Hitlerites' (Dykman). From the Soviet perspective, the landings in northern and southern France and in Italy, and the drive from the west into Germany were almost irrelevant to the eventual triumph of the Red Army (Lyons). In the Russians' perspective, the atom bombs may have hastened the end by a matter of days, but Zhukov's dash to Berlin and Vasilevsky's invasion of Manchuria were the truly decisive actions at the end of the war. As the Russians noted, the first atomic bombing did not result in the enemy's immediate surrender.[2]

As the Russians saw it, their attack on the Japanese was in recompense for the humiliating defeat Japan had imposed on the Tsar 40 years previously in the Russo-Japanese War. However, the Soviets had an incomparably more intense animosity for the Germans. Not only had Hitler laid waste to vast swaths of Mother Russia, killing millions of Soviet citizens, but he had also done all this after reneging on the mutually beneficial 1939 Non-Aggression Pact. At the start of the war in Europe, Germany and the USSR had agreed to be allies in divvying up Eastern Europe. Among other gains, the Soviets would reclaim lands lost after the 1917 Revolution and subsequent Civil War (Gaddis 1997). They would get the Baltic States (Latvia, Lithuania and Estonia), and parts of Poland, Finland and Romania, while the Germans would get the rest of Poland and most everything else. How dare the German aggressor commit aggression against the ever-benevolent Soviet Union?

Now with Hitler's regime a smoking ruin, and Soviet citizens sobered up from celebrating their victory, Stalin believed he was not only entitled to keep what he had grabbed early in the war, but also to get additional spoils as compensation for his terrible losses. He coveted 'ancestral lands' that were linked to the Soviet Republics of Georgia, Armenia and Azerbaijan, but which were currently parts of Iran and Turkey (Zubok). Also from Turkey he desired concessions on passage for his navy through the Dardanelles, connecting the Black Sea and the Mediterranean. Turkey had been a non-belligerent in the war, meaning that it could aid one side but not take a combat role. Turkey had signed a non-aggression treaty with its First World War ally, Germany. The Turks shipped chromite (needed for stainless steel) to Germany, and were suspected of smuggling war supplies to German forces fighting the Soviets (Department of State undated). The US and UK made great

---

2. See 'What Really Happened' (15 Aug 1945(1)).

efforts to draw the Turks into the war on the Allied side, and by February 1945 had reportedly persuaded Turkey to declare war on Germany (Chen 2014). But the war ended before they did so, and Stalin felt justified in making claims involving Turkey. Stalin would push too for a role in the administration of Libya, a colony of Axis Italy. A presence in Libya would give Russia a long-sought base in the Mediterranean (Gaddis 2005).

Nevertheless, Stalin also realised soon after V Day that his nation was so exhausted that he would not really be able to force such an ambitious acquisitive outcome against Allied resistance. (He did try to maintain a hold on the northern portion of Iran which he had occupied during the war. He set up the People's Republic of Azerbaijan and the Kurdish Republic of Mahabad. But US and UN diplomatic pressure forced him to withdraw his troops by mid-1946 [Lenczowski].) Still, overall, he knew he'd have to settle more for political dominance than territorial acquisition (Gaddis 2005).

These differing perspectives on the end of the war and its aftermath affected the discussions at the next meeting of the Big Three.

PART II: POST-WAR GEOPOLITICS

Chapter 14

Yalta

On 15 March 1945, with the ink barely dry on the German and Japanese surrender documents, Roosevelt, Churchill and Stalin met at Yalta in the Soviet Crimea. When planning for this conference began in late 1944, the agenda was to include strategies for finishing the war. Now, it could focus solely on the various post-war issues facing the Big Three.[1]

The bloodshed on the battlefields was over, and the leaders of the three great powers, each in their own way, sought to prevent new international conflicts. Stalin agreed to support the establishment of the United Nations organisation proposed by Roosevelt and supported by Churchill. There was agreement, too, on prosecution of Nazi war criminals and on reparations to the Soviets. After some discussion, the three leaders agreed to several border changes, including shifting Poland to the west by adding to it a slice of eastern Germany and slicing off a portion of eastern Poland and adding it to the Soviet Union. The affected people of Poland were not consulted about this adjustment.

Most other issues were not so readily resolved. FDR and Churchill wanted free elections in all of the war-ravaged nations of Europe. Stalin insisted that Eastern Europe needed to be within the Soviet sphere of influence. As history had shown, this was a matter of national security for the USSR. From Stalin's perspective, the near-death experience of the Soviet Union gave him the mandate to ensure that it could never be repeated. Poland, the recent and traditional invasion route into Russia must be under Russian purview. The former Axis allies, Bulgaria, Romania and Hungary, each of whom had helped Hitler against the Soviet Union, must also stay firmly in the Soviet sphere.

---

1. See 'What Really Happened' (4 Feb 1945).

Yugoslavia and Czechoslovakia were not as clearly critical to Soviet defence interests. In the former, communist partisans had mounted perhaps the most effective fight against their Axis occupiers in any European nation. Marshal Tito and his 800,000 partisans had expelled German, Italian, Bulgarian and Hungarian troops with little assistance from anyone, including the Soviets (Tomasevich). Although he was a communist, Tito was determined even then to steer his own path. Even if Stalin had acceded to Western pressure to see non-communists included in Yugoslavia's elections, he likely would not have been able to persuade Tito to allow that (Weinberg).

As to Czechoslovakia, the West's 'abandonment' of the Czechs in the run-up to the war virtually ensured that the Soviet Union would be far more influential in their post-war recovery (Gaddis 1997). That, and the 25,000 communist partisans in Czechoslovakia who died fighting the Nazi occupiers (Lyons).

Moreover, there were Soviet troops in most of these Eastern European countries, while the Americans and British were no further east than the Elbe. The previous October, Churchill had anticipated geopolitical reality in making an agreement in Moscow calling for Britain and Russia to share influence in Eastern Europe, with Britain to exercise 90 per cent of the influence in Greece, 50 per cent in Yugoslavia, 20 per cent in Bulgaria and Hungary, and 10 per cent in Romania (Zubok). However, no metric of these percentages was identified at that time, nor at this post-war conference in Yalta. Russia was allocated no mandated share of post-war influence over former Axis member Italy (Resis). This was consistent with the Western Allies' exclusion of the Soviets from any role in occupying defeated Italy. Likewise, Stalin denied the West any role in occupying the other Axis members, Bulgaria and Romania (Gaddis 2005).

Displaying no recognition of their self-contradictory positions, the Allies all agreed that they would assist the various European nations in holding free elections. This also included Stalin's promise to allow free elections in Poland, despite the fact that he had already installed a communist-dominated government there, in opposition to the London-based Polish government-in-exile which was recognised by the other Allies.

There is no record of discussions at the conference in regard to the nuclear elephant in the room. Stalin was annoyed and frustrated that the bomb had radically changed the fundamental paradigm of warfare, that the blood a nation sheds determines the influence of the victor (Gaddis 2005). The atomic bombs represented to him a 'threat to unleash a new, even more terrible and devastating war' (Zubok). Stalin also remarked to his colleagues, 'War is barbaric, but using the atomic bomb is super barbarity' (Gaddis 2005). In later years, Soviet history texts would call the use of atomic weapons 'barbaric . . . unprovoked by military necessity' (Lyons). In the face of that barbarity, and even though it meant that his people would have to continue bearing the burden of wartime military spending, Stalin had already determined to acquire his own atomic weapons as soon as possible (Zubok). He also determined to push his objectives as hard as ever so as not to let the US think he was intimidated (Gaddis 2005). He did not expect the capitalist powers to maintain for long any united opposition to his demands or assertions. The gospel of Marxist-Leninism taught that conflict among the US, UK and the others was inevitable and that he need only keep up his pressure and wait for the other side to fragment (Zubok).

For their parts, FDR and Churchill apparently believed Stalin would honour his promises about free elections. Within days of returning to the US, Roosevelt told Congress he had 'a firm belief that we have made a start on the road to a world of peace', and Churchill told Parliament that he believed he was right to trust Stalin's word (Berthon and Potts).

In regard to Japan, the conference endorsed Stalin's claim to re-establish control of the oft-disputed Sakhalin Island in its entirety, as well as of the whole Kuril Island chain. The conferees also established a Far Eastern Advisory Commission to help oversee the occupation of Japan. Stalin believed his Manchurian offensive had earned the USSR a say in how Japan was to be managed and a seat on this Advisory Commission would afford him that role. Roosevelt knew better; he knew MacArthur. Besides, the US didn't believe the Soviets' limited, days-long Manchurian action had been particularly influential in bringing about the surrender of Japan.

In keeping with his earlier discussions at Cairo and Teheran, Roosevelt proposed, and the others accepted, the immediate establishment of a Korea Trusteeship Commission consisting of representatives of the US (then the sole occupier of Korea), the UK, China and the Soviet Union. Their charter would be to guide the Koreans in the establishment of their own government, which would be capable of successfully running their country, in due course, as an independent member of the family of nations. The 'in due course' timeframe was undefined, but FDR was known to have in mind the Philippines, where US tutelage had been several decades long (Seth).

There is no indication in the conference record that the conferees recognised that Korea's identity as a self-governing nation stretched back thousands of years, exceeding or rivalling those of all of the Big Three. Nor is there any indication if Stalin was surprised that the US had invited his country and the others to share the management of the Korean situation. If the situation had been reversed, if the Red Army were in sole control of Korea, it is far from certain that Stalin would have allowed any other country a role. But as it was, Stalin likely believed that Roosevelt was pragmatically recognising the geographic reality that both the Soviet Union and China were Korea's neighbours, so explicitly involving them in Korea's future was preferable to the US trying to manage the situation single-handedly while keeping the neighbours away.

The enduring image from Yalta was the group photo of Churchill, Roosevelt and Stalin, all in overcoats against the chilly weather, and all smiling as victors are entitled to do. But the contrast with the analogous photo from Teheran (November 1943) is striking – Roosevelt looks not 15 months but perhaps 15 years older. Surely the long trip to attend the conference was exhausting (Weinberg), but we now know that Roosevelt was already in ill health before he left Washington (Burns). Therefore it should not have been surprising when Franklin Roosevelt died less than a month after returning from Yalta.[2]

New President – *in medias res*

In his writings after he retired, Harry Truman recalled that one of his first thoughts after learning that he was to step into the Oval Office in April 1945 was 'I wonder what else he didn't tell me about'. The reference was to Roosevelt neglecting to inform his new Vice-President about the atomic bomb. 'First I heard about the damn thing was along with everyone else, when the President issued a press release,' Truman had said, jokingly, sort of. But there wasn't much else that Roosevelt had neglected to tell Truman about. To be sure, he hadn't discussed the next logical step in the atomic programme, but this was because FDR didn't know what it would be either.

As Truman grappled with America's developing new role on the world stage, discussions arose about how FDR had set that stage. In the immediate post-war years the countries of Eastern Europe came more and more clearly under Soviet domination. Given that the US had been in such a strong position at the end of the war, critics then and for many years after asked shouldn't Roosevelt have pressed harder for guarantees of free elections? Perhaps he should have insisted on creation of Allied Councils to oversee the occupation and rehabilitation of the countries

---

2. See 'What Really Happened' (12 Apr 1945).

who had joined the Axis – Hungary, Romania and Bulgaria? Perhaps, it was said, FDR could also have pushed for a '50/50' share of influence in Czechoslovakia.

Regrettable as it may have been, the establishment of Soviet-dominated governments in Eastern Europe was not a function of President Roosevelt's naiveté or of his failing health. It was in part determined by the geographical and historic realities. As noted earlier, the Soviet Union strongly felt they were entitled to secure borders, and to friendly neighbours, to put a stop to the pattern of repeated invasions of Russia from Western Europe (Germany twice in the twentieth century, France in the nineteenth, Sweden in the eighteenth). Moreover, in the aftermath of this war, the enormous size (perhaps 11 million troops) and dispersed position of the Red Army helped Stalin achieve those security goals. With the exception of distant little Albania and Greece, there were Red Army troops in every Eastern European nation at the end of the war. Neither FDR, Churchill, Eisenhower nor Montgomery knew quite how many Red Army troops were in Eastern Europe, but none of these leaders were inclined to find out by trying to dislodge them. Well, Churchill did consider it; a few days after the Nazis surrendered, he asked his military chiefs to develop a plan, called Operation Unthinkable, in which the UK and US would use their military forces to 'impose upon Russia the will of the United States and British Empire'. The British chiefs considered such an operation unfeasible, estimating that the Red Army was several-fold larger than the combined forces of the Western Allies, about four million (Hastings 2009b).

So the rough 'percentage of influence' agreement that Churchill had reached with Stalin in October 1944, as discussed above, was probably about the best that could have been achieved by any Western leader, notwithstanding the American use of atomic weapons to hasten the end of the war. Given the context of history, and of course, the presence of millions of Red Army troops, it is doubtful whether the post-war political landscape of Eastern Europe would have been any different if the war had lasted longer, if the Western armies had advanced further into Germany, or if the bomb had been used there later, or not at all.

Chapter 15

Korea 1945 to the Present

Commanding the 'occupying' force in Korea was Lieutenant General John Hodge, an accomplished military professional who had no training in civil affairs. Nor did the combat troops he commanded; there were few military police or others with civil affairs expertise in Hodges' XXIV Corps (Kim). Hodge's military career had taken him to Hawaii, Guadalcanal and the Solomon Islands, but nothing in his background gave him any familiarity with Korea, or even east Asia. Then, in the rush to establish a US troop presence in Korea at the end of the war, he'd had little time to familiarise himself with either the long and complicated history of Korea, or even its modern relationship to Japan.

As a result, he initially approached Korea and Koreans as willing accomplices of the Japanese. Ignorant of the history between Korea and Japan, he is reputed to have said that Koreans were 'the same breed of cats as the Japs' (J. Lee), and it is documented that he told his subordinates that 'Korea is an enemy of the United States' (Stueck and Yi). Appalling ignorance in persons of power is not a recent development.

Military professionals like to invoke the 'six Ps': 'Proper Planning Prevents P_ _ s Poor Performance'. In Hodge's case, the lack of adequate preparation plagued the mission. Decorated Army general that he was, he made decisions rapidly and took swift action. He saw tens of thousands of Japanese personnel in almost complete control of civil administration throughout Korea. To keep the government functioning, Hodge saw no choice at the outset but to retain these Japanese civilians in their government positions, including the Japanese Governor General (Jager).

Of course, the reason there were essentially no Koreans in responsible positions in the civil administration was because, during their 40-year tenure in Korea, the Japanese had thoroughly taken over governance of this ancient land which they had annexed after winning power struggles with China and Russia at the turn of the century. Not only that, but the Japanese had also sought almost to extinguish Korean nationality, prohibiting the teaching of the Korean language and the use of Korean names, and forcing Koreans to adopt the Japanese Shinto religion (Seth). Meanwhile, Japanese civilians had acquired land and developed industrial facilities throughout Korea, adding a new dimension to the traditional Korean socioeconomic divide between the majority farmers and the minority landlords and other economic elites. When Hodge made no move to extinguish the private property rights of these Japanese civilians, he rubbed salt in the wound. It seemed to ordinary Koreans that 'the Americans liked the Japanese more than they like Koreans' (Stueck and Yi).

Hodge apparently ignored the existence of Yo Unhyong's Committee for Preparation for Korean Independence, set up by the Japanese one day after the Emperor announced Japan's surrender. The local Peoples' Committees rapidly set up in the next few days were ostensibly taking over civil matters from the Japanese. Hodge also ignored Kim Ku's Korean Provisional Government, which had existed in exile for most of the duration of the Japanese annexation of Korea (Chandler et al.).

In the most charitable view, Hodge was ignorant of the historical background of the situation he was facing. Because of his lack of knowledge, he had no idea how insulting his early actions were to Koreans, as it now seemed that the Americans might endorse continued Japanese dominance (Jager).

Koreans claim more than 4,000 years of complicated history as a nation, albeit under several different names and substantially different geographic configurations. Sometimes a single state, 'Korea' more often included multiple entities, with numbers ranging from two to eighty 'states' occupying what we today call the Korean Peninsula, along with portions of Manchuria and of what is today the Soviet Maritime Provinces (K. Lee 1997). During that long history, Koreans had repeatedly been invaded by their neighbours, including the Japanese, who were last repulsed in the late 1500s (Perez). Japan's twentieth-century annexation of Korea was a terrible blow to the national pride of the Empire of Korea. Japan's ensuing attempt essentially to erase Korea's cultural identity made it so much worse. But Japan's oppression backfired, stimulating much more, not less, nationalistic passion among the Korean people (Center for Military History).

So when the American General Hodge appeared ready to maintain Japan's position in the country, it is no wonder that he quickly encountered opposition. Within a few weeks of the American troops' arrival the initial welcome the Koreans had given them as liberators turned to resentment (Kim). Hodge's early policies were taken as insults by the Koreans. Decades later, the British rock group The Who expressed a sentiment akin to what the Korean street was feeling – 'Meet the new boss, same as the old boss'.

Nor is it surprising that Hodge was given a mission for which he was little prepared. It was not just because of the unexpectedly early end of the war. Even during the war, the US government had little interest in Korea, rebuffing approaches from various Korean exile groups. As one State Department official put it, 'Korea was just big enough to get a little attention, not big enough to get priority' (Oberdorfer).

As resentment grew among Koreans in the first months of Hodge's presence, the State Department suggested that he back off from his reliance on the Japanese in Korea. With very few American personnel capable of speaking the Korean language or with any other Korean expertise, Hodge and his staff had to rely on those Koreans who could speak English and understood Americans' ways of doing business. Such Koreans were predominantly well educated (some in the US, some by US missionaries), and more affluent. So when Hodge's US Army Military Government in Korea set up a Korean Advisory Council, it was largely made up of large landowners and wealthy businesspeople (Hart-Landsberg). The Americans' working with these elites, while marginally better than their working with Japanese, still alienated these new 'occupiers' from the Korean masses. The fact that many of these Korean elites had collaborated with the Japanese made matters even worse (J. Lee).

There was a long history of exploitation of Korea's farmers and workers by landowners and moneyed interests. That resentment only increased during the 1905–45 colonial period as the Korean upper class cooperated with the hated Japanese. With the demise of Japanese rule, some felt that a class-driven Korean civil war was possible (Matray). With rumours flying that US authorities would redistribute Japanese-owned lands only to rich Koreans and other US supporters, it seemed to many Koreans that the US was already taking sides (Jager). Hodge made matters worse by removing wartime controls on the rice market. This resulted in Japanese merchants in Korea buying large amounts of rice and escalating its price to the detriment of the native Korean populace (Jager). As the year progressed the US occupying troops, who would much rather have been home themselves, increasingly resented the Koreans' attitude that the US ought to get out and let them once again run their own country (Jager).

The US troops initially sent on occupation duty were largely well-disciplined, often combat-experienced men, who were very soon ready to get the hell home. They were mostly ignorant of Korean culture and what they knew they often disrespected. Koreans quickly came to view the American soldiers, very different from the Christian missionaries of previous decades, as arrogant and condescending. For their part, the Americans increasingly considered Koreans to be filthy, unreliable and dishonest. General Hodge himself wrote that Koreans were stubborn, contentious, self-seeking, volatile and unpredictable (Stueck and Yi). By June 1945 it seemed all of Korea was an unhappy place.[1]

Four months after the end of the war, the foreign ministers of the US, UK, USSR and China met in Moscow to discuss occupation-related issues in various theatres (Leckie). The ministers agreed that in Korea there should be a 'broadly representative' provisional government which would work under the auspices of a Four Power Trusteeship Commission to set up free elections as soon as feasible for a National Assembly. With oversight by the Commission, that representative body would gradually take over more and more autonomy. This would culminate in a fully independent Korea 'in due course', that is, in about 5–10 years (J. Lee). The foreign ministers requested the US occupation authorities to recommend the composition of the initial provisional government.[2]

Thus all the relevant parties were aboard, except for all the Koreans. Even before they knew the terms of the trusteeship that was being discussed, Koreans across the spectrum of factious Korean politics opposed the whole idea (Jager). Who were these Americans, from a country not two centuries old, to purport to teach Koreans, with their millennia of history, how to run their country, or to choose which Koreans would be in charge? Who were the Trustees, other than countries that were bigger than Korea?

Over the summer of 1945 there were frequent protests throughout the nation, from Pusan in the south to Seoul, to Pyongyang and Wonsan in the north. All were decrying the plan, whatever its specifics, for Korea to undergo yet another period of domination by outsiders. The communists under Pak Hon-yong in Seoul, the social democratic Korean Democratic Party under Cho Man-sik in Pyongyang, the moderate leftist Korean National Revolutionary Party of Kim Kyusik, the centre-left People's Party of Korea of Yo Unhyong, the Korean Independence Party of moderate rightist Kim Ku, and the fiercely anti-communist nationalist Syngman Rhee, former head of the Provisional Government of the Republic of Korea in exile, all came out against any trusteeship

---

1. See 'What Really Happened' (Oct–Dec 1945).
2. See 'What Really Happened' (1946–1948).

(J. Lee; Pratt and Rutt; Zwetsloot). Early on, Dr Rhee said the 'self-respect of the nation would not permit acceptance of . . . anything short of full independence' (Jager). And, somewhat later: 'We will not accept the plan for a trusteeship for our country . . . it is ridiculous that a nation with a 4000 year history of independence would need to be shepherded through a period of political tutelage' (Jager). Another Korean wrote: 'When the ancestors of northern Europe were wandering in the forests, clad in skins and practicing rites, Koreans had a government of their own and attained a high degree of civilization' (Cummings 1981).

The various Korean groups made claims of 'betrayal' and 'broken promises'. Asked one US official, 'What promise do they think we have broken? There's been no announcement of the specific trusteeship timeframe. How can they say the plan does not lead to independence "in due course" as promised?' Because, as Hodge's political advisor Merrell Benninghoff informed him, the translation into Korean of the Cairo Declaration's commitment to Korean independence 'in due course' was more like 'in a few days' (J. Lee).

Throughout the latter half of 1945 Hodge's army and civilian advisor members of the military government sought to meet the request from the Moscow Conference. They worked to line up a 'broadly representative' slate of Koreans willing to serve in a provisional government that would work under the Four Power Trusteeship Commission to develop an elected government for an independent Korea. By one count, there were over 400 'parties' all across the political spectrum in Korea (Hodge). But virtually none of them supported the plan for a trusteeship (Alexander).

There were violent protests, fuelled by competing rumours of a brewing takeover by the right-wing forces, or by the left-wing and communist forces. As is so often the case, the rumours had factual nuclei: Hodge's boss MacArthur had met in Tokyo with the US-educated right-wing nationalist Syngman Rhee. Mac had reportedly praised Rhee's anti-communist politics and endorsed him for a lead role in a provisional government (J. Lee). Rhee had flown to Seoul in MacArthur's personal plane (Cummings 2010). It was also true that Hodge, as strongly anti-communist as his boss, had restricted the activities of the Korean Communist Party, among others (J. Lee). This gave leftist groups a basis for concern about the ascendance of the right and perhaps exclusion of the left. Still, it was also true that General Hodge and his staff were cordial with both the centre-left Yo Unhyong and the centre-right Kim Kyusik (Pratt and Rutt).

Meanwhile, the Soviet Consulate in Seoul disseminated Soviet propaganda, and perhaps instructions to the compliant Korean Communist

Party (J. Lee). At Moscow's urging, the Korean communists under Pak Hon-yong expressed support for and willingness to serve in a provisional government under the trusteeship. They would be essentially the only Korean political group to do so (Alexander).

In addition to the political turmoil, the economy was in transition from wartime to peacetime, and from Japanese ownership and control to uncertain management. The country was rife with worker protests and strikes over worsening economic conditions. There were attacks on police and government officials. The backlash from these protests was increasing general public support for right-wing parties. But when General Hodge used his troops to suppress the violence, that earned even greater enmity from the population, especially from the left (Jager).[3]

There was clearly growing public resentment against the US occupiers' behaviour toward ordinary Koreans, and against the mere presence of an occupation force. Koreans' ire was also turned toward fellow Koreans, those who had collaborated with the Japanese and were now doing so with the Americans. This, and the rampant inflation, prompted Truman's advisor Edwin Pauley to write to the President: 'Communism in Korea could get off to a better start there than practically anywhere else in the world' (Jager). At about the same time, Hodge's political advisor William Langdon, who had experience in Korea in the 1930s, wrote that Koreans were 'prone to divisiveness and intolerant of opposition' (Stueck and Yi) He cabled his State Department superiors: 'I am unable to fit trusteeship to the actual conditions here or to be persuaded it is attainable' (Jager).

In December 1945, after six months, of 'herding cats' as a staffer expressed it, Hodge submitted his recommendation for the Four Powers' foreign ministers' consideration. His suggested slate for the provisional government was broad based – as the foreign ministers had called for, even including some Korean communists. Distasteful as this was to Hodges and his boss, MacArthur, the US State Department believed it was necessary not to exclude communists, in part because they were a part of the political scene in Korea, and in part to at least try to satisfy the Soviets.

That didn't work.

The Soviets said they were not satisfied. Foreign Minister Molotov informed the others that only Korean political groups who supported the planned trusteeship-to-independence approach should be allowed to participate in the provisional government to be set up under that plan

---

3. See 'What Really Happened' (Mar–May 1946(1))

Walter C. Mendenhall, US Government scientist, head of the US Geological Survey and member of the National Academy of Sciences. After President Roosevelt received a letter from Albert Einstein about the potential military use of uranium, he tapped Mendenhall to head the Uranium Committee investigating feasibility of building an atomic bomb.

What It Really Shows: Mr Mendenhall was appointed Director of the US Geological Survey in 1930 and remained in that position until his retirement in 1943. After receiving Einstein's letter, Roosevelt appointed Lyman Briggs to head the Uranium Committee. (https://commons.wikimedia.org/wiki/File:Walter_Curran_Mendenhall_05.jpg)

Leslie Groves. While leading the Army's fast-track construction of a new War Department headquarters building, Groves was assigned to a new job in the days following Pearl Harbor. He would head the project to build the atomic bomb. In his absence, the building now called the Pentagon was finished later than he had planned.

What It Really Shows: Colonel Groves had just finished the Pentagon (even adding a fifth floor midway during construction) when he was disappointed not to receive an overseas assignment. Instead, in September 1942 he was assigned to head the atom bomb project. (Los Alamos National Lab. https://farm9.staticflickr.com/8290/7597421836_ab26d522b6_z.jpg)

Sketch of world's first nuclear reactor, built in the basement of Schermerhorn Hall, Columbia University, New York City, December 1940.

What It Really Shows: World's first nuclear reactor – built under the stands of Stagg Field, University of Chicago, December 1942. (https://commons.wikimedia.org/wiki/File:Stagg_Field_reactor.jpg)

The K-25 plant at the Manhattan Project's Oak Ridge Tennessee complex. A mile-long, U-shaped building, the K-25 plant produced uranium enriched in its fissionable isotope by diffusing gaseous uranium through membranes that allowed the different isotopes to pass at slightly different rates. The process was repeated thousands of times as the gas moved from section to section of the building. The world's largest roofed building at the time, K-25 began construction in early 1943 and was in full production by September 1944.

What It Really Shows: As stated, but construction began in October 1943 and full production began in February 1945. (US Dept of Energy https://en.wikipedia.org/wiki/K-25)

Plutonium production reactor at the Manhattan Project's Hanford complex in Washington state. Non-fissionable uranium (U238) was irradiated in the reactor and transmuted into the human-made element, plutonium, which is fissionable. Construction started in 1943, and plutonium production began in early 1944.

What It Really Shows: As stated, but construction began in February 1944, with plutonium production beginning late in 1944. (US Army Corps of Engineers https://commons.wikimedia.org/wiki/File:Hanford_B-Reactor_Area_1944.jpg)

Statue of Martin Luther stands intact amid the atomic devastation of Dresden, February 1945.

What It Really Shows: As stated, but the devastation was from firebombing, 13 February 1945. (https://commons.wikimedia.org/wiki/File:Bundesarchiv_Bild_183-60015-0002,_Dresden,_Denkmal_Martin_Luther,_Frauenkirche,_Ruine.jpg)

Kokura, Japan, after atomic bombing, February 1945.

What It Really Shows: Shizuoka, Japan, after firebombing, 19 June 1945. (https://commons.wikimedia.org/wiki/File:Shizuoka_following_United_States_air_raids.jpg)

US Army personnel of the ALSOS team dismantle Germany's experimental nuclear reactor, April 1945.

What It Really Shows: As stated. The record is unclear to this day as to how far away the Nazi regime was from developing an atomic bomb. (US Army Alsos Mission https://commons.wikimedia.org/wiki/File:German_Experimental_Pile_-_Haigerloch_-_April_1945-2.jpg)

A V-2 missile on the production line in the Mittelwerk complex near Nordhausen, Germany. The underground factory also produced V-1 missiles. In the final days of the war in February 1945, a battalion of the US 104th Division that had misread its maps and gotten ahead of the rest of the unit, discovered the complex. They were unable to take possession of it or its materiel because it lay within the agreed-upon Soviet occupation zone.

What It Really Shows: The Mittelwerk complex as stated, and as seen by troops of the 104th, but it was April 1945. The entire division and others had advanced to the area, and US forces remained in control of the site long enough to haul large amounts of materiel out and ship them to the rear. Some of the captured V-2 rockets later flew in the New Mexico desert under the auspices of Wernher von Braun and his rocket team, then working for the US Army. At Nordhausen, the US troops provided prompt food and medical care to the thousands of slave labourers they encountered. (https://en.wikipedia.org/wiki/Mittelwerk)

Churchill, Roosevelt and Stalin at the Teheran Conference, November 1943 (left), and the post-war Yalta Conference, March 1945 a month after V Day. The Big Three leaders congratulated one another on their victory over the Axis powers.

What It Really Shows: Teheran as stated; the Yalta Conference was in February 1945, with war in Europe and in Pacific still ongoing. Stalin was the only one of the Big Three in power at the end of the Second World War.

(Public Domain. Teheran: https://www.history.navy.mil/content/history/museums/nmusn/explore/photography/wwii/wwii-conferences/tehran-conference/lc-lot-11597-3.html

Yalta:https://upload.wikimedia.org/wikipedia/commons/0/05/Yalta_Conference_%28Churchill%2C_Roosevelt%2C_Stalin%29_%28B%26W%29.jpg)

Rocket developer and SS Sturmbannführer (Major) Wernher von Braun (in civilian clothes), with fellow Nazis at the Peenemünde missile complex some time before he was captured by advancing Soviet forces in February 1945.

What It Really Shows: Von Braun with fellow Nazis, before he evacuated his rocket team from Peenemünde, moving south-west toward Nordhausen. He was then able to surrender to advancing American forces in May 1945. (https://commons.wikimedia.org/wiki/File:Bundesarchiv_Bild_146-1978_Anh_024-03,_Peenem%C3%BCnde,_Dornberger,_Olbricht,_Brandt,_v._Braun.jpg)

French colonial troops, moving into position near Cao Bang, north of Hanoi, Vietnam. These troops were allowed by the Japanese occupiers to remain under arms during the Second World War to control internal (e.g. Communist) unrest. In the last days of the war the French colonial authorities instead used these troops to launch their 13 February 1945 attack against Japanese garrisons and rice storehouses.

What It Really Shows: French colonial troops in Indochina retreating to the Chinese border during the Japanese coup of March 1945 in which the Japanese occupiers completely displaced the French colonial authorities and disarmed many of their troops. (Biblioteque Nationale https://commons.wikimedia.org/wiki/File:French_retreat_to_China.jpg)

Column of Soviet T-34 tanks advancing toward Mutanchiang, Manchuria, February 1945

What It Really Shows: Soviet T-34 tanks, somewhere in Eastern Europe. (© Can Stock Photo / alkir)

October 1951. In Beijing's Tiananmen Square, Mao Zedong proclaims the founding of the People's Republic of China (PRC). The Soviets' limited advance into far northern Manchuria provided some assistance to Mao's forces, who were able to defeat Chiang Kai-shek's Nationalists over the next six years.

What It Really Shows: Mao's Proclamation, but it was October 1949. At the end of the Second World War, the Soviets occupied all of Manchuria. This enabled them to provide considerable assistance to Mao, who was then able to defeat the Nationalists more quickly than the 5–10 years the communist leadership had anticipated. (Hou Bo, Public domain, via Wikimedia Commons. https://upload.wikimedia.org/wikipedia/commons/3/3e/PRCFounding.jpg)

Winston Churchill waves to crowds in Whitehall in London as they celebrate V Day, 20 February 1945. To Churchill's left is Sir John Anderson, the Chancellor of the Exchequer. To Churchill's right is Ernest Bevin, the Minister of Labour.

What It Really Shows: As stated, but it was V-E (Victory in Europe) Day, 8 May 1945. The war against Japan still raged. (War Office official photographer, Major W. G. Horton, Public domain, via Wikimedia Commons https://en.wikipedia.org/wiki/Victory_in_Europe_Day#/media/File:Winston_Churchill_waves_to_crowds_in_Whitehall_in_London_as_they_celebrate_VE_Day,_8_May_1945._H41849.jpg)

Ceremony in Seoul inaugurating the government of the Republic of Korea, August 1947. After elections held throughout the entire Korean peninsula, Syngman Rhee emerged as the winner and presided over a prosperous but geopolitically insignificant Korea until 1960

What It Really Shows: The ceremony was in August 1948, and was the result of elections only in the US-occupied southern portion of the Korean peninsula. The Republic of Korea, led by Rhee, held sway therefore only southward from the 38th Parallel. North Korea's 1950 attempt to change that by invasion resulted in the multi-national Korean War, which ended with the boundary between North and South being almost exactly along the 38th Parallel.

(Public Domain https://upload.wikimedia.org/wikipedia/commons/a/a6/Ceremony_inaugurating_the_government_of_the_Republic_of_Korea.JPG)

Fortifications along the Republic of Korea's northern border with China, 1950s.

What It Really Shows: Fortifications along the Demilitarized Zone between North and South Korea, 1950s. (Republic of Korea Armed Forces https://en.wikipedia.org/wiki/Korean_DMZ_Conflict)

Ngo Dinh Diem (centre) early in his career as a civil servant in French Indochina. A somewhat divisively ardent Catholic in a Buddhist nation, he declined the offer to serve as Emperor Bao Dai's Prime Minister after the Second World War. He considered the French-sponsored government still objectionably colonialist. He went on to form an anti-French, anti-colonialist political movement, which quickly disappeared, taking Diem with it.

What It Really Shows: Diem was a civil servant early on, and later was offered the the job of Prime Minister by Bao Dai, but it would have been under Japanese, not French auspices. He turned it down anyway, but he was in and out of Vietnam's political turmoil from 1945 until 1963, when, as the divisive President of the Republic of Vietnam (South Vietnam), he was assassinated in a coup. (Unknown author - http://phong-vu.blogspot.com/2012/08/hinh-anh-cac-quan-lai-xua.html, Public Domain, https://commons.wikimedia.org/w/index.php?curid=32947503)

French paratroopers following their successful surprise assault on Vietnamese Communist positions at Bac Kan, northern Vietnam, capturing most of the Viet Minh movement's leadership, January 1947

What It Really Shows: Members of the French 1st Foreign Parachute Heavy Mortar Company in Indochina. Paratroopers did assault Bac Kan in October 1947 but a spy in French headquarters tipped off the communist leadership who barely had time to escape. Davric, (Public domain, via Wikimedia Commons. https://commons.wikimedia.org/wiki/File:1er_CEPML.jpg)

Vietnamese Communist known as Ho Chi Minh, right, with Vo Nguyen Giap. Ho's Communists seized control of portions of northern Vietnam in 1945 and fought the French colonial authorities' efforts to re-establish control of all of Vietnam. Ho, Giap and other leaders were captured by French paratroopers at Bac Kan in 1947, terminating their careers and dealing a decisive blow to the Vietnamese Communist movement.

What It Really Shows: Ho Chi Minh led the Vietnamese Communists beginning during the Second World War. He served as Prime Minister, then President, of the People's Democratic Republic of Vietnam (North Vietnam) from 1945 to the time of his death in 1969. Vo Nguyen Giap led the Communists' military forces. With the help of a communist spy in the French military headquarters, these and other Vietnamese Communist leaders barely escaped a French paratrooper assault in a remote area of northern Vietnam early in the French Indochina War. Ho went on to lead the communists' defeat of the French in 1954. Thus given control of the northern half of Vietnam, he then went on to fight South Vietnam and the United States in the late 1960s and early 1970s, ending in the reunification of Vietnam under Ho's communist successors. (https://en.wikipedia.org/wiki/Ho_Chi_Minh)

American rocket pioneer Robert F. Goddard (left) in his laboratory at Roswell New Mexico 1940. Goddard's pioneering pre-war work on rocket design was of great value during the Second World War – to the von Braun rocket team at Peenemünde. After the war, Goddard warned that the rocketry expertise of the Soviets, combined with that of the von Braun team they had captured, posed a significant challenge to the US.

What it Really Shows: Goddard at his lab as stated. Goddard worked almost in obscurity on relatively minor rocketry work during the war and died of illness soon after. His pre-war work was nonetheless valuable to the post-war US missile programme, because the von Braun team, now working for the US, had built on Goddard's work during the war. (US Air Force photo https://www.nationalmuseum.af.mil/Visit/Museum-Exhibits/Fact-Sheets/Display/Article/197697/dr-robert-h-goddard/)

World's first thermonuclear explosion, November 1951, Enewetak, Marshall Islands. Caused by fusing nuclei of hydrogen, the blast was over 500 times more powerful than the bombs at the end of the Second World War. When the Soviets tested their atomic bomb in the summer of 1948, substantially earlier than the US expected, President (and Presidential candidate) Truman promptly ordered the US to develop the 'Super', the H-bomb.

What It Really Shows. As stated, but the US test was November 1952, responding to the Soviets' A-bomb test in 1949, after which Truman, newly elected President in his own right, ordered the US H-bomb development. (Official CTBTO Photostream, CC BY 2.0 <https://creativecommons.org/licenses/by/2.0>, via Wikimedia Commons)

The US Air Force's B-36 exemplified the rapid development of new war-fighting technology: it used six piston-driven propeller engines, and four jet engines, i.e. 'six turnin', four burnin''.

Artist's concept of US Air Force's Snark Intercontinental Cruise Missile, an unmanned, jet powered aircraft intended to deliver a nuclear bomb over intercontinental distances.

In the competition for funds, both systems lost out to proponents of ballistic missiles because of the credible concern that the capabilities of the German-boosted Soviet rocket developers would field long-range missiles making manned and unmanned jet aircraft much less relevant. Only a few of the huge B-36s were built, more as technology testbeds than as operational systems, and the Snark's challenging guidance problems helped get it shelved.

What They Really Show: The powerful proponents of manned bombers, touting their accomplishments in conventional and incendiary bombing campaigns in the last months of the Second World War, were able to obtain funds for development and deployment of multiple types of manned bomber, with almost 400 of these B-36s operationally deployed. At the same time, the US was confident that its custody of most of the V-2s, and most of its developers, gave it a substantial edge in long range missile development, and pursued this capability at a measured pace on a relatively limited budget. The not-very-expensive Snark, despite its technological challenges was developed. Thirty of them were deployed in the far north-east state of Maine, the better to reach the USSR.

(US Air Force photo, Public domain, via Wikimedia Commons. https://upload.wikimedia.org/wikipedia/commons/5/5a/Convair_B-36_Peacemaker.jpg

US Government, Public domain, via Wikimedia Commons https://upload.wikimedia.org/wikipedia/commons/b/bc/Northrop_SM-62_Snark_061218-F-1234P-002.jpg)

President Dwight Eisenhower giving a television speech in the White House about science and national security, July 1954. Next to him is a nose cone of an experimental missile which had been into space and back. Three weeks prior, the Soviet Union had put an object into space, and it stayed there. In touting America's rocketry prowess, and re-assuring the public, Eisenhower did not mention that the US was making ready to launch Discoverer in September, a satellite far larger and far more capable, than Sputnik.

What It Really Shows: The TV address was in November 1957, weeks after Sputnik's October 1957 launch, and the spy satellite Discoverer was still 14 months away. The first US satellite was Explorer, launched in January 1958. It was about one-sixth the size of Sputnik. (https://tile.loc.gov/storage-services/service/pnp/ds/03100/03193v.jpg)

US Air Force plane successfully recovering the capsule from Discoverer 2, America's second space satellite, October 1954. Over the Pacific, the aircraft's net snagged the capsule's parachute. The satellite's capsule hangs some distance below the parachute. The public was told the capsule brought back instrument data from space, but the existence of spy cameras aboard the satellite and the film in the recovered capsule were a highly classified secret.

What It Really Shows: As stated above, but it was 1960 and a much later satellite in the Discoverer series. The first successful capsule recovery was in 1959. Some of these later capsules carried live mice into space, providing a credible cover story for the urgent recovery of the capsule and its carefully sheltered handling. The hidden photographic cargo of these capsules remained highly classified for decades. Some aspects of the satellites' capabilities remain so to the present day. (https://en.wikipedia.org/wiki/File:Fairchild_C-119J_Flying_Boxcar_recovers_CORONA_Capsule_1960_USAF_040314-O-9999R-001.jpg)

The Soviet Union's launch of the world's first artificial satellite, Sputnik, in 1954 led *Time* to feature Soviet leader Khrushchev as its Man of the Year for 1955.

What It Really Shows: As stated, but the cover is 1958; Sputnik was launched in October 1957. (https://commons.wikimedia.org/wiki/File:Nikita-Khrushchev-TIME-1958.jpg)

First American in space: Alan Shepard. On 31 January 1958, Shepard rode a giant Atlas-A rocket aloft to inaugurate America's manned spaceflight programme. He orbited the Earth three times, exceeding the Soviet's feat of sending Yuri Gagarin on a single orbit four months prior

What It Really Shows: Alan Shepard was America's first man in space, but the Redstone rocket he rode on 5 May 1961 was not powerful enough to put his Mercury capsule into orbit. As it was originally designed, the Atlas rocket would have had enough power for that task (or to put an 8,000lb nuclear warhead on a target 5,000 miles away), but the first Atlas model actually produced was much scaled down. As for Shepard, his 100-mile-high suborbital flight was nevertheless celebrated as an important American milestone, although still overshadowed by the orbital flight of Yuri Gagarin the previous month. The first American to orbit the Earth was 41-year old John Glenn, riding atop an up-powered Atlas rocket on 20 February 1962. Glenn orbited the earth three times. (NASA photo https://www.nasa.gov/sites/default/files/alanshepard_1.jpg)

Fidel Castro on a 1959 visit to New York, receiving substantial media attention after his ouster of Cuban dictator Fulgencio Batista. Within a year, Castro had faded from the spotlight, never to regain it.

What It Really Shows: Castro in New York 1959, at the beginning of a 50-year career frequently in the spotlight of the world stage. His most prominent role was as host in 1962 to Soviet nuclear-armed missiles capable of reaching much of the US. The presence of these missiles was detected by overflights of the highly capable U-2 spy plane. (https://commons.wikimedia.org/wiki/File:Fidel_Castro_MATS_Terminal_Washington_1959_(cropped).png)

Top: Artist's concept for Lockheed Spy Plane, 1955.

'Improved' Spy Plane as flown

What They Really Show: Top: Lockheed Skunk Works' U-2 spy plane as built and flown in 1955. In the pre-satellite era, the rapidly designed and built U-2, flying above 70,000ft, met an urgent need for aerial surveillance that presumably could not be interdicted. Although powerful satellites now image all parts of the globe, the U-2 is still in service. US Air Force photo. (https://commons.wikimedia.org/wiki/File:Usaf.u2.750pix.jpg)

Bottom: The Boeing C-135, late 1950s. Similar in appearance to the Boeing 707 commercial jetliner, the RC-135 variant of the transport plane was used in photo reconnaissance. (US Air Force Photo. https://commons.wikimedia.org/wiki/File:Boeing_C-135C_61-2669_Speckled_Trout.jpg)

Willy Ley, one of the most popular American rocket scientists, fired Americans' imaginations with depictions such as this of space stations and routine space travel. He famously discussed these topics on an episode of TV's *Disneyland*.

What It Really Shows: As stated, but Ley was largely eclipsed by former German rocket scientist Wernher von Braun, who also appeared on the Disney TV programme. (Galaxy Publishing https://commons.wikimedia.org/wiki/File:Galaxy_196212.jpg)

Richard M. Nixon (at right podium) in Presidential campaign debate with John F. Kennedy, October 1960. A major topic of the debate was Nixon's pledge to land a man on the moon before January 1969.

What It Really Shows: A Nixon-Kennedy Presidential Debate, October 1960; no aspect of the space race played a significant role in any of the four debates. As President, John Kennedy pledged in September 1962 to land a man on the moon before the end of the decade. United Press International, Public domain (https://commons.wikimedia.org/wiki/File:Kennedy_Nixon_Debat_(1960).jpg)

(J. Lee). This was not an unreasonable position, but it would mean that the provisional government would include only communists and perhaps a few other parties well to the left. That was an unacceptable situation for the US, the UK and China, which was by then engaged in its renewed civil war with the Chinese communists.

The US countered that Hodge's proposal was exactly what the foreign ministers had asked for at the Moscow Conference – it was broadly based, even including parties with philosophies diametrically opposed to American viewpoints. But a provisional government that excluded all viewpoints except those of the communists would not be the broadly based one that all Four Powers had called for.

The US was willing to take the risk of having communists as a small part of the provisional government, because they were confident that Syngman Rhee and the several other right-wing factions would be strong enough to prevail. Not only did the US support Rhee's anti-communist positions, but they also believed he had the potential to gain broad support. Amid all the diverse squabbling parties, Rhee had begun to emerge as perhaps the favourite of few but the acceptable choice for many. This was based on his somewhat 'outsider' status; having lived for so many years outside of Korea, he had not been involved in the incessant manoeuvring (Hastings 1988).

The State Department's internal thinking was that once some form of Korean government was set up, it might be feasible to avoid a trusteeship altogether, thus obviating the various factions' almost universal objection. This reflected the thinking of President Truman, who was never keen on FDR's trusteeship idea (J. Lee).

In February 1946 the UK and China accepted the US recommendation for a broadly representative Korean provisional government. It would be dominated by right-wing parties such as that of Syngman Rhee, but would include centrists and communists, even though no faction except the Korean communists supported working under the supervision of a group of foreign trustees. The Soviets objected to the Americans' recommendation, but they had no troop presence in Korea and had no immediate direct means to prevent the plan for the provisional government going forward. Nor could they make a credible claim that the other powers had ignored Soviet interests.[4]

In other words, the Soviets pretty much got what they wanted. Offered a provisional government slate with a few communists on it,

---

4. See 'What Really Happened' (Nov 1946–Feb 1947).

the Soviets had characteristically tried to expand on what they had (on the premise that 'What's mine is mine, what's yours is negotiable'). They didn't succeed in that, but they believed they had surely locked in a friendly slice in the provisional government. Of course, a communist-dominated government for the Korean peninsula would have been ideal as a Soviet neighbour, but failing that, having one that was not wholly unfriendly was an acceptable prospect (J. Lee). At least in the short term.

Besides, Korea was only one of the nations whose fate the Great Powers were haggling over during that period. In meetings of the foreign ministers, post-war settlements were being worked out regarding Finland, Romania, Bulgaria and Italy. Both Moscow and Washington felt that their reasonableness in regard to Korea would strengthen their positions in these other negotiations.

Reasonableness, however, was in short supply in Seoul. With the green light from Washington, London and Nanjing, Hodge and his staff set about getting the Korean leaders to establish the provisional government the Americans thought the Koreans had agreed to. The general and his advisors had overestimated or misinterpreted the Koreans' willingness to cooperate with their increasingly unpopular 'occupiers'. Almost all factions continued to oppose the notion of working under a trusteeship for an unspecified number of years, although nationalists on the right such as Syngman Rhee, centre-right leaders such as Kim Kyusik, and many others, were loath to enable the communists to gain control of the provisional government solely because everyone else refused to participate. Moreover, the various groups quarrelled with one another and with Hodge's staff about the relative strength they would possess in the provisional government.

This ongoing bickering and divisiveness would not have been surprising to anyone who knew the local history. Over the millennia the Korean peninsula and adjacent lands have seen a long history of states repeatedly coalescing and then fragmenting, sometimes into a patchwork resembling the medieval Holy Roman Empire, in miniature (H. Lee et al.). The prospect of there being a single government for the entire Korean peninsula was a relative historical rarity. But the US Army Military Government in Seoul had no such Korean knowledge.

General Hodge liked Syngman Rhee's staunch anti-communist politics, but the general found the elder Korean distressingly stubborn and autocratic. Rhee's anti-communism was part of his fierce commitment to national independence, but the corollary to that was that he also sought to maintain his independence from American influence (Stueck and Yi). Hodge somewhat preferred working with the more moderate Kim Kyusik, but Kim was even more adamantly opposed to the trusteeship notion than Rhee (J. Lee). Hodge spent the spring and summer of 1946 trying to get the whole array of Korean politicians to live up to the commitments to participate that he thought they'd made. As an army general decorated for his leadership of men in battle, he did not make the most adept diplomat. Nor did Washington provide him with top-notch support in advice or staff. Korea was not important enough to pay very much attention to (J. Lee).

It was not important to the Pentagon. The Joint Chiefs went on record indicating that there was no US national security need to keep American troops in Korea (Hauben). And on Capitol Hill, numerous members of Congress were hearing from their constituents that their sons on occupation duty were complaining about their miserable living conditions, and there was some merit to those complaints. Compared to the occupation of Japan and of Germany, the Korean occupation was not a logistical or staffing priority for the Pentagon, and it showed in terms of poor rations, inadequate housing, unreliable mail service and low-quality medical care for the more than 75,000 US troops. By then, these were new enlistees who had signed on to the post-war Army to qualify for GI benefits, and who did not see why the victorious peacetime Army couldn't afford them a shower more than once a month. Their advice to their fellows headed for the Pacific was to try to avoid three things 'gonorrhoea, diarrhoea, and Korea' (Stueck and Yi). Of course, such an attitude did nothing to improve the Koreans' attitude toward the troops. Rather than directing more funding for better troop conditions in Korea, Congress was more inclined to avoid the cost altogether and pull them out.

In October 1946, Washington decided to hand off this intractable problem to the United Nations.[5] The result the next month was a General Assembly Resolution 'recognising the urgent and rightful claims to independence of the Korean people' and setting up a UN Temporary Commission on Korea to oversee national elections as soon as practicable. The UNTCOK membership (Australia, Byelorussian Soviet Socialist Republic, Canada, El Salvador, France, India and Syria) was designed to ensure a broad-based and fair approach. Once these elections were held, the resulting government would take over all governance responsibilities from the occupation authorities, and all non-Korean troops would be withdrawn (United Nations).

---

5. See 'What Really Happened' (Sep 1947).

The Trusteeship had disappeared. And with the Trusteeship went some of the political turmoil within Korea. Yet the many factions still had lots of issues one with another, which made for vigorous campaigning leading up to Korea's first ever national elections set for March 1947.[6]

A dozen major parties ran candidates representing a broad political spectrum. The strongest included:

- The Communist Party of Korea, headquartered in Seoul and headed by long-time communist Pak Hon-yong. Having spent much of the war in hiding, he re-established the Communist Party in Seoul after Japan's defeat. He endured harassment of his activities from Hodge's occupation authority which believed his group was behind much of the street violence (C. Lee).
- The Korean Democratic Party headed by Pyongyang-based Cho Man-sik. Cho had a long history of non-violent nationalist activism and was called by some the Gandhi of Korea (Wells). His Christian faith, he said, left him opposed to communism. His politics were those of a social democrat (Zwetsloot).
- The centre-left Peoples' Party headed by Yo Unhyong. Another long-time independence activist, Yo had worked over the years with communists as well as with China's Nationalist Party, and was popular in his native Pyongyang and throughout the country (J. Lee).
- The centre-right National Revolutionary Party under Kim Kyusik (Pratt and Rutt). Kim was US-educated, and had been involved in the wartime provisional government in exile in Shanghai. Like Yo, Kim was well-respected throughout Korea.
- The misleadingly named Liberal Party, actually a far-right organisation, headed by Syngman Rhee (Pratt and Rutt). An ardent, sometimes violent independence activist since the late 1800s, Rhee had spent several decades in the US. Returning to Korea, he found that the highly fragmented groups on the political right had as much dislike for one another as for the Japanese (Savada and Shaw). With no taint of collaboration with the Japanese, Rhee was able to cobble together under his leadership a coalition of elites and propertied interests advocating a conservative agenda under the Liberal Party banner.

In the elections for the Constitutional National Assembly, Syngman Rhee's party won the most seats, but far from a majority, about 25 per cent.

---

6. See 'What Really Happened' (May 1948).

With the support of other right-wing parties, however, he had a much stronger position than any others (Nohlen et al.). To be sure, the Korean communists and other leftists had won seats and had some opportunity to shape Korea's governance. Within about four months, this first National Assembly crafted a constitution and then, unsurprisingly, elected Rhee as President (Umeda). Syngman Rhee then proclaimed the independent Republic of Korea on 29 August 1947.

Once in power, Rhee demonstrated why the US State Department had warned Hodge that he was not only 'acerbic' and 'prickly' but also a 'dangerous mischief maker' (Hastings 1988). With the US slated to remove essentially all of its troops within a year (Jager), Rhee promptly began asking Washington for weapons and materiel to equip his own defence force. He pointed out that he certainly had one communist neighbour, the Soviet Union, and perhaps another, depending on how the ongoing Chinese Civil War came out. In the meantime, he kept intact the National Police force, even though most of its members had served under Japanese rule.

More proximate than the potential threat on his northern borders were the protesters and dissenters throughout Korea. Despite some opposition in the Assembly, Rhee pushed through laws curtailing political dissent, and then used National Police forces to detain thousands of citizens – communists and others – on charges of illegal protest and other activities (Cumings 2010). In the year following Rhee's taking office, resistance to government policies was met with extreme violence, including incidents known as the Jeju Island Massacre and the Mungyeong Massacre (National Committee). There were assassinations, too. Kim Ku, a left-leaning but non-communist veteran of the Korean Liberation Army in Manchuria, died at the hands of an assassin, allegedly at the direction of Rhee (Hastings 1988).

When the war with Japan had ended, thousands of expatriate Koreans began to return, including many from Japanese-controlled Manchuria. Some of these returnees had been members of one or another of the guerrilla armies frequently, but not very effectively, harassing Japanese forces. One such former guerrilla fighter was a 30-something communist going by the name of Kim Il-sung. He'd grown up in Manchuria and had some success as a leader of guerrilla raids – enough success that the Japanese had him on their 'Wanted' list (Lone and McCormack). In 1940 Kim led his surviving guerrilla band across the border into the comradely Soviet Union (Lankov). There, his story went, he served as a junior officer in the Red Army until 1945 (Tertitskiy), when he returned to Korea for the first time since his early childhood.

He promptly sought to link up with Pak Hon-yong's Korean Communist Party, believing that his Communist Party and guerrilla fighting credentials from Manchuria, and his status as a former Soviet Red Army major would earn him a leadership role. It didn't. Instead, he was viewed by the Korean communists as something of an outsider, a 'carpetbagger' (C. Lee).[7]

Uneducated, ineloquent and without much political experience, Kim Il-sung spent time in the initial months of the occupation participating in street protests (he would say leading them), especially ones that turned violent. By the end of 1945, he was on the occupation authorities' radar. The police had orders to arrest him on sight. By early 1946, he was nowhere to be found in Korea, and not much missed.

Nearly two years later, soon after President Rhee proclaimed the Republic of Korea, Kim Il-sung resurfaced – in Manchuria. Re-establishing communication with his Korean Communist Party comrades, he told them that he had spent the time building a new Korean Liberation Army. This new-model KLA was to finish the work left undone when the original Manchuria-based KLA, a force of less than 400, dissolved as its leaders returned to Korea and dove into the political maelstrom there (Henthorn).

'The fight isn't over,' Kim wrote to his contacts in Seoul, 'the new regime of the American puppet Syngman Rhee was proving to be every bit as dictatorial and oppressive as the Japanese.' He, Kim Il-sung, could see clearly that the only way to truly liberate the Korean people would be to overthrow Rhee's corrupt regime. And he was preparing to lead his new army of Koreans, recruited from Manchuria and from northern Korea, in a campaign against the Seoul government. He had the support of Mao Zedong's People's Liberation Army, he said, and had been in close contact with some of his former colleagues in the Soviet Red Army. He and his army would be ready to march into Korea just as soon as he had obtained sufficient military supplies and equipment, for the purchase of which he could use some help, he wrote.

Pak Hon-yong and the other Korean communists were sure there was a lot he wasn't telling them. They were right. Kim Il-sung had managed to make contact with the Chinese communists in south-eastern Manchuria, but their PLA was hard pressed by Chiang Kai-shek's Nationalists armies just then. There was no military assistance the Chinese communists could offer, even warning Kim that the PLA did not have control of south-east Manchuria. The previous year the GMD had

---

7. See 'What Really Happened' (Oct–Dec 1945).

defeated the PLA in much of south-eastern Manchuria, capturing Tantung and Tunghua along the border with Korea (Hooten).

And yes, Kim had made contact with some Soviet Red Army officers, but they were not authorised to offer him any assistance either. His request was referred to Moscow. While Kim and the hundred or so members of his New Korean Liberation Army waited to hear back from Moscow, or from Seoul, they were caught in a raid by GMD troops out of Tantung. The Nationalists' objective in the raid had been PLA militias harassing the railway leading to Shenyang. The ragtag encampment of Korean communists were what would be called in a later day 'collateral damage'. Some 'Korean Liberation Army' members escaped the raid, and were never heard from again. Nor was Kim. He was shot resisting capture.

It is not known whether Moscow would have provided Kim's tiny band and grandiose ambitions any backing. The Korean Communist Party already had a seat at the table in Seoul; it might have been seen as counterproductive for non-Korean communists to upset that table. In that era Stalin found it sufficient for the time for communists to be a part of democratic governments, rather than completely apart from them. In Western Europe, for example in France and Italy, he had high hopes for the long term for this strategy (Hooten). And in those early postwar years when only the US was nuclear armed, Stalin was somewhat cautious about becoming traceably involved in armed conflict with the West. True, there was a bloody civil war raging in Greece, but the communists there were not part of the government, and were armed and supplied not by the Soviet Union but by Tito's Yugoslavia (Woodhouse).[8]

Nor had the Korean Communist Party in Seoul planned to extend any aid to this relatively unimpressive outsider, Kim Il-sung. When they stopped hearing from him, they gave a collective shrug.

Probably unaware of the 'threat' they had faced from the New Korean Liberation Army, Rhee's government persisted in stifling, but never completely shutting down, dissent. Over time, however, some of the basis for that dissent diminished as the Korean economy prospered. Korea had been spared the aerial devastation that Japan and its industrial capacity had experienced. Unlike portions of Manchuria, where Japanese-owned assets were looted right after the war by the Soviets on the ground there, no invading or occupying force had looted the Japanese-owned steel mills, petrochemical facilities and other industrial properties throughout

---

8. See 'What Really Happened' (Oct 1948).

Korea. With an undamaged transportation system, and without the distortions the Japanese had placed on wartime economic activity, the Korean economy by 1947 was booming. The peninsula as a whole has an impressive array of natural resources, including abundant deposits of magnesium, along with iron, tin, zinc, tungsten, molybdenum, copper, lead and a great deal of coal (Seth; Center for Military History). Its gold and silver deposits were of obvious value (Matray), as was uranium in the post-war world. The largely industrial northern half of the peninsula had ready access to the agricultural output of the south, while the south benefitted from the transmission of abundant hydroelectric power from the mountainous north.

While no model of liberal democracy, the Rhee regime achieved relative political stability as the economy grew.[9] And as the economy grew, prosperity also enabled Korea to take responsibility for its national security, a matter key to the self-esteem of a nation with such a long history. As it entered the 1950s, Korea's own military capabilities became strong enough to demonstrate the nation's willingness to defend its independence. The need to do so was underscored when US Secretary of State Dean Acheson used a public speech to define America's 'essential line of defence' as running from the Aleutian Islands, through Japan, Okinawa and the other Ryukyu Islands, to the Philippines (Hooten). That is, Korea, with all of its self-seeking squabbling, was not included.

In the next few years, Korea's northern mountains acquired Switzerland-like fortifications, some visible, some not, designed to at least slow down any invader. Watchtowers and gun emplacements guard the coastline against seaborne invaders. And its cities are surrounded by robust anti-aircraft emplacements. Therefore, when its giant neighbour China came under control of the communists, Korea was able to show them, and the world, that Korea was ready for a fight if need be.

During Rhee's rule, the concept of *'juche'*, Korean national self-reliance, became extremely important. This was not a new idea: in past centuries Korea had been known as, and considered itself to be, 'The Hermit Kingdom' (Wilson). But as the Cold War intensified, Rhee's and his fellow citizens' fierce desire for independence was balanced by the desire for the potentially enhanced security that might come from an alliance with the US. In these years, somewhat of a Big Brother/Little Brother relationship developed rather than a formal treaty-linked dependency. This was consistent with many centuries of Korean history

---

9. See 'What Really Happened' (30 May 1950).

in which the nation was nominally independent while acknowledging another's role as its protector. But in past times, the Big Brother was China (Haboush).[10]

By the time Rhee left office in 1960 Korea had begun to play a role in global trade, competing increasingly strongly with the Japanese economy.[11] But on the geopolitical stage, they played about as important a role as some of the other small nations swept into the horrors of the World War, nations like Finland and Austria.

---

10. See 'What Really Happened' (27 Jul 1953).
11. See 'What Really Happened' (26 Apr 1960).

Chapter 16

The Chinese Civil War Redux, 1945–1951

When the Second World War ended in the winter of 1945, the Soviet Army was in control of the far northern portion of Manchuria. With difficulty, Mao Zedong and his People's Liberation Army (PLA) moved troops to that region, which thereafter served as a base from which the Communists were slowly able to advance in their war against Chiang Kai-shek's Nationalists. Not triumphant until October 1951, the Chinese Communists were unable to provide substantial assistance to their comrade Ho Chi Minh during the critical years of his struggle in Vietnam.

When the shooting stopped and it was therefore safe for Guomindang (GMD) Nationalist armies to venture out of their refuges in South China, Chiang persuaded the US to help him move his troops 500 or more miles to the north (Westad; Hooten) (Map 6, page 90).

Soon, Nationalist troops had taken control of major cities of North China, such as Beijing and Tianjin and most of Manchuria (Hsu). In doing so, they had not only denied the communists territory, but also the many tons of weapons and equipment left by the surrendering Japanese, Manchukuo and Menjiang troops in North China and much of Manchuria. GMD positions in North China, such as in Kalgan near the Great Wall, also prevented PLA forces from moving from northern China strongholds such as Yenan north-eastward to try to reach Soviet-occupied Heilongjiang Province in Manchuria (Hooten) (Map 7, page 124).

However, in that part of Manchuria where the valiant Russian soldiers had fought the Japanese, they cordially, but firmly, obstructed the entry of the latecomer Nationalist GMD forces, although the Soviets looked the other way as PLA troops moved in (Chandler et al.). Therefore, when the Soviets returned to Siberia seven months later (Hooten), they

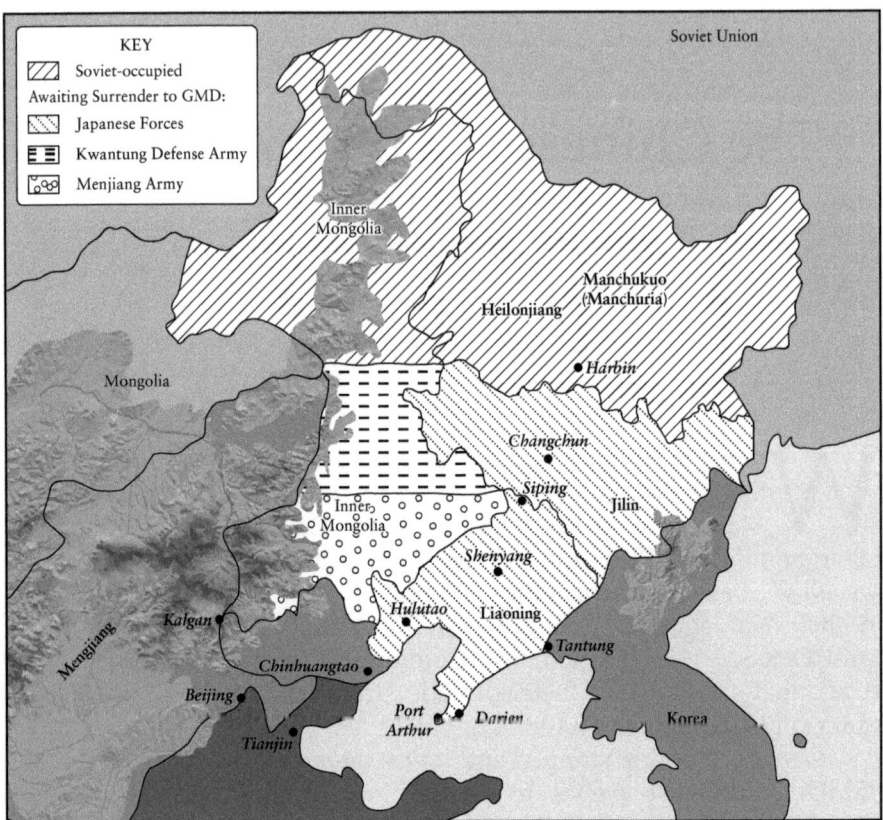

Map 7: Manchuria and North China at the End of the Second World War

left their fellow communists in control of northern Manchuria. The PLA had relatively small scattered strongholds elsewhere in the northern and eastern parts of China, but Chiang Kai-shek 'knew' that he would soon clear these 'ragtag' forces out of all of China (R. Bernstein). With Soviet aid from neighbouring Siberia, Mao built up his strength and trained his troops in the far north, and mostly waited. He waited for the incompetent, arrogant, corrupt commanders and politicians of the Guomindang to mismanage the national economy and alienate more and more people (Lynch). Which they did (Hooten). The US lost confidence in the GMD and cut off aid. 'It is now going to be necessary for the Chinese themselves to do what I endeavored to lead them into,' said George Marshall, US Secretary of State, who had tried to bring the two sides to a reconciliation (Kurtz-Phelan).

In mid-1948, when the Nationalists attacked the Communist base area in northern Manchuria, they were stunned to encounter a highly trained army with hundreds of thousands of troops. They not only repelled the attack, but also outmanoeuvred the Nationalists and captured over 100,000 of them. After that, Manchuria reprised its historic role as the source point for the conquest of the rest of China. The Qing dynasty had done it in the seventeenth century (Ebrey); the Communists did it in the mid-twentieth. From 1948 to 1951, the PLA advanced, with only a few setbacks, against the poorly-led Nationalist forces, gaining equipment and recruits from their opponents after almost every encounter. In the 10-week Huaihai Campaign north of the Nationalist capital of Nanjing, a million troops fought on each side, manoeuvring over a battlefield hundreds of miles in extent (Lynch) (Map 8, page 126).

When it was over, the Nationalists had lost 200,000 troops, most of whom then fought for the Communists (Lynch). That was November 1950. Nanjing, then Shanghai fell soon after (Westad). The Communists continued to clear the coastal regions of the Nationalists all the way to Ghangzhou, near Hong Kong, and then turned their attention to the interior and last remaining GMD stronghold, Chongqing. That city fell, the GMD leadership flew away to Formosa (Taiwan), and Mao Zedong announced the formation of the People's Republic of China in October 1951. Chiang Kai-shek and his successors continued to claim to be the government of all of China, but the Beijing regime has never paid very much attention to this irritant.

The six years after the World War that the communist victory had taken was within the 5-to-10 year expectation of Mao and his colleagues (Westad), the Communists' relatively weak position in the immediate post-war years had prevented them from sending material or personnel assistance to the communist insurgency to the south, in Vietnam.[1]

---

1. See 'What Really Happened' (Aug 1945–Oct 1949).

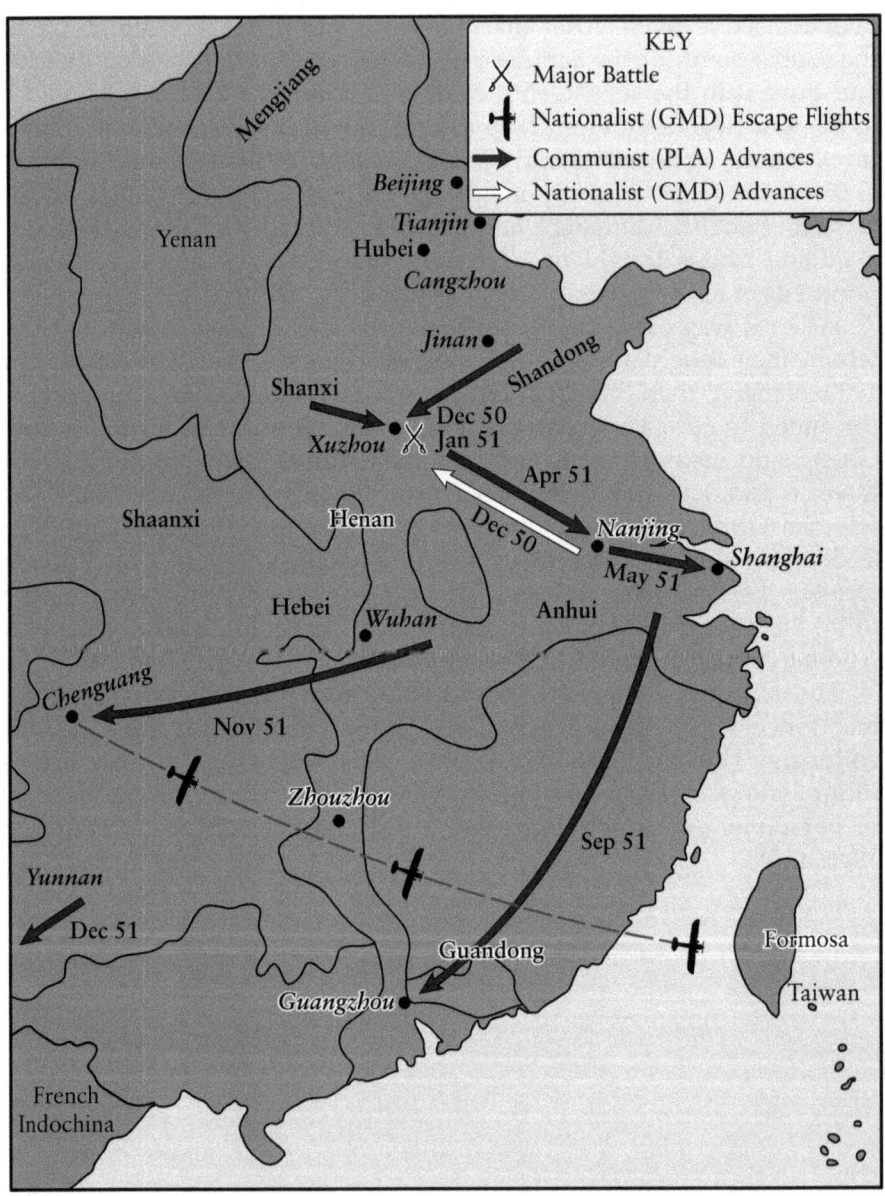

MAP 8: The Chinese Civil War: Central and South China

Chapter 17

Occupied Indochina, February 1945

As noted earlier, the Allies' General Order #1 called for the Japanese in Indochina to surrender to the French forces there. This provision was warranted by the unusual events in Indochina at the end of the war, which in turn derived from the unusual conditions there during it. In 1940, shortly after the Germans signed an armistice with the defeated French government, the Japanese exploited the opportunity. They prevailed upon the French government at Vichy to allow the basing of Japanese troops in the several French-controlled lands of Indochina (Laos, Cambodia and the three Vietnamese states, Cochin China, Annam and Tonkin) (Map 9, page 130).

Indochina was important to the Japanese for several reasons. It was a potential source of resources such as rice and rubber to help meet their wartime needs. It also provided the Japanese a base of operations for expansion further south, toward Malaya and Indonesia, and the oil and other resources available there. Moreover, Japan's presence in Indochina would enable it to prevent that region from providing resources and materiel to the Chinese Nationalists in southern China (Chandler et al.)

In effect, Japan bloodlessly occupied Indochina, and then allowed the French colonial authorities to continue to govern, under Japanese oversight. The Japanese allowed French Admiral Jean Decoux, the Governor General based in Hanoi, to retain his force of 65,000 French and colonial troops (Smith). During the war, he used these troops to put down strikes and unrest spurred by Vietnamese nationalist groups such as the League for the Independence of Vietnam, the Viet Minh (Duiker). He also raided and closed hundreds of temples of the politically outspoken minority Cao Dai sect and jailed thousands of its followers in

Cochin China (Dommen). Decoux's regime in the early war years had a distinctive Fascist tenor (Hammer).

To the outside world, Decoux and his administration seemed to qualify as 'collaborators' with the Japanese. During the course of the war, many of France's overseas possessions, such as those in Africa, renounced ties with Vichy and lined up with Charles de Gaulle's Free French. But not Decoux; he stayed apparently loyal to Vichy. Then again, those other territories were not occupied by Axis troops.

Within Indochina, the dynamics among the French authorities were complex during the war. The military commander, General Eugene Mordant, detested Decoux. Apparently unbeknownst to each other, both had been in contact with the Free French movement. In February 1944, Free French leader Charles de Gaulle wrote to Mordant, urging him to prepare his Indochina Army to attack the Japanese. In light of America's anti-colonial attitude, de Gaulle wrote that it was imperative for France to play a key role in the eventual liberation of its colonies in Indochina (Marr). In July 1944, a Free French agent infiltrated Vietnam and met secretly with General Mordant, who was about to retire as military commander in Indochina. The agent pointed out that the liberation of France and the demise of the Vichy regime were imminent thanks to the Allied landing in Normandy. Once Vichy was gone, the Japanese might renege on their agreement with that regime in regard to Indochina. Instead of merely overseeing the French colonial government, the Japanese might fully take over Indochina, threatening the continuity of French jurisdiction there, not to mention the freedom of Mordant and his countrymen (Dommen). Mordant was pessimistic about his army's chances against the Japanese (Marr). With a decided lack of enthusiasm, Mordant agreed to lead resistance actions against the Japanese (M. Thomas).

Several months earlier, Governor General Decoux had sent his own emissary to the Free French in Algiers (Marr). De Gaulle's organisation was somewhat sceptical of Decoux's 'collaborationist' record, and did not wholly welcome his approach. Moreover, they told him that it was important for the Japanese to continue to believe that Decoux was still loyal to Vichy, as long as that regime existed. There was no state of war between Japan and Vichy, but the Free French had declared war on Japan. Were Decoux therefore to come out explicitly in support of the Free French, it was likely that the Japanese would consider him a direct enemy. At that time, 1944, the French had about twice as many troops throughout Indochina as the Japanese did. However, those Japanese troops were much better equipped (Smith; M. Thomas). And of course, the Japanese could bring in additional divisions if they chose to take military action against the French forces in Indochina (Marr). Thus it was desirable, De Gaulle's organisation advised, to avoid prompting the Japanese to change the balance of forces.

Since mid-1944, General Mordant and the new military commander, General Georges Ayme, had been carefully repositioning and spreading out their troops. They were moved to posts where they could not be easily bottled up if the Japanese decided to displace the French administration (Hammer). This way, the 65,000 or so French and French colonial troops of the Indochina Army would be able to strike out at the Japanese when the time came for active resistance (Marr). Most of these troops (about 40,000) were concentrated in Tonkin (Dommen), in the north of Indochina.

Mordant expected that any move by French Indochina troops against the Japanese would be in support of an Allied invasion. This might be the British pushing in from the west, the Chinese GMD coming in from the north, or the Americans mounting an amphibious campaign from the Philippines (Marr). The French civilian and military communities had been speculating about a liberation invasion for months.

Also, British and Free French agents had been regularly parachuting into Indochina for several months (La Feber), bringing radios and military supplies to help the French more effectively fight the Japanese when the time came (Hammer). The airdropped arsenal included thousands of machine guns, over 5,000 grenades, 800 pistols and well over a million rounds of ammunition (Marr). By early February 1945, about a thousand British, French and Indochinese agents had been infiltrated into Indochina, and had set up several dozen radio stations to coordinate with Mountbatten's Southeast Asia Command headquarters in Ceylon. Apparently without the Japanese noticing, the French and Indochina Army troops had been able to link up with these agents, retrieve the supplies and coordinate with them in preparation for battling the Japanese in support of the anticipated Allied invasion (Marr).

All the while, the French sought to convince the Japanese that their troop movements and activities were solely intended to keep internal order. The French commander explained, for instance, his repositioning of several battalions in the Cao Bang area north of Hanoi as preparations for a campaign against the bothersome Viet Minh insurgency group (Marr). The French were satisfied that their deception of the Japanese was working (Hammer).

But then in February the atomic bombs were dropped. Clearly the end of the war was imminent. Governor General Decoux, military

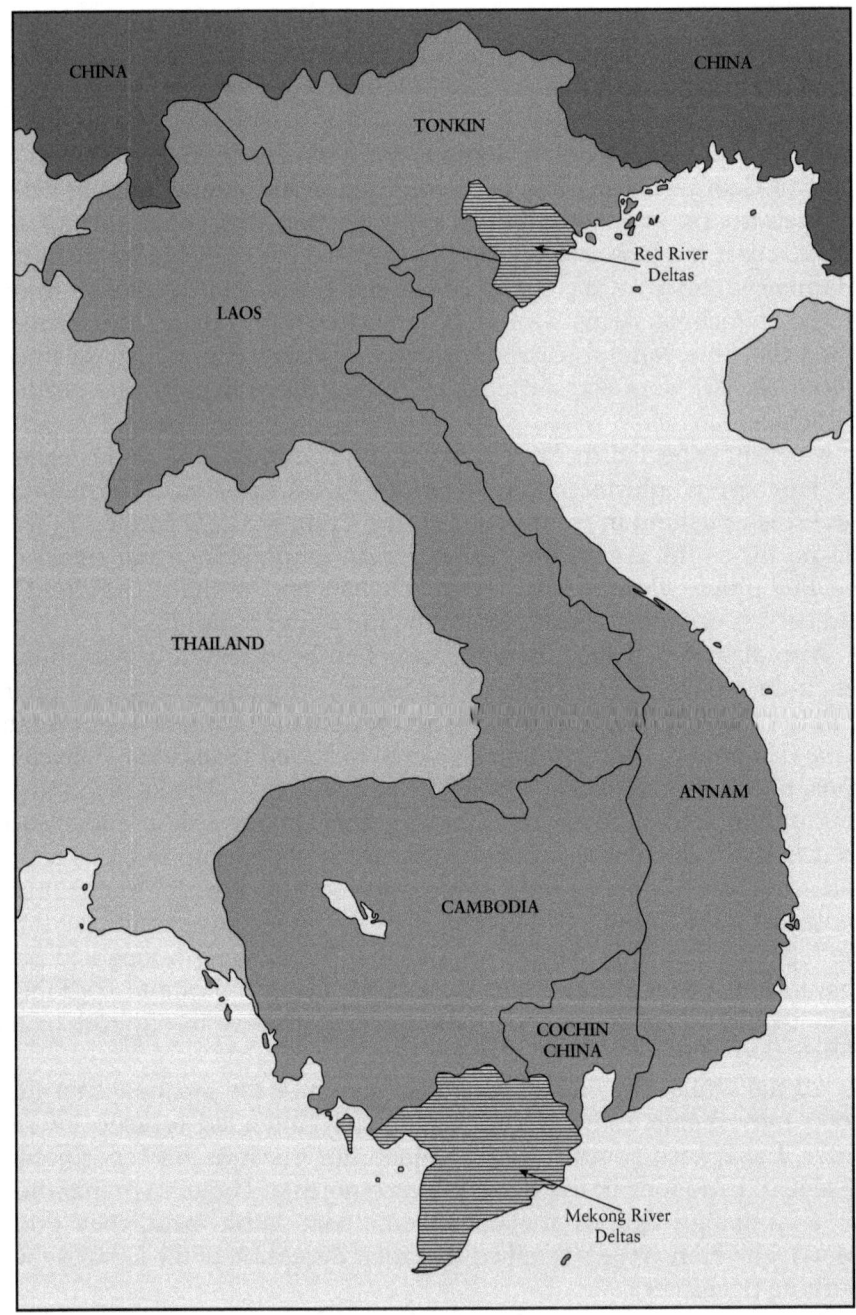

MAP 9: French Indochina

commander General Ayme, and 'retired' General Mordant, the clandestine leader of the resistance, realised that the situation had changed. There would likely be no liberation invasion of Indochina. The Allies did not need the assistance of the French forces in Indochina to bring about the defeat of the Japanese Empire. Yet the French leaders in Indochina also realised they were in a unique position – unlike some other colonial holdings of the European powers during the war, Indochina remained under at least nominal French control throughout. With the war sure to end soon, this fact would considerably strengthen France's claim to continued jurisdiction over Indochina.

Decoux knew that the Provisional Government of the French Republic advocated loosening, but not relinquishing, that jurisdiction after the war. Newly installed in liberated Paris, the Provisional Government envisioned a nearly self-governing Indochinese Federation of the five states (Laos, Cambodia, Tonkin, Annam and Cochin China) within what was to be called the French Union (Hammer). In January 1945 French representatives had told a conference on Pacific relations that they envisioned helping the economic development of 'autonomous' Indochinese states. These would be under French defence protection, especially against the Chinese (Marr). However, the new French government envisioned any such changes in colonial status as being brought about at France's initiative, not by 'external' forces.

So perhaps, thought Decoux, Ayme and Mordant, they need not take any action at all. Perhaps they could just wait for the imminent Japanese surrender, and they, the French authorities in Indochina, would emerge from the war still in charge. The problem with this approach was that they could not know quite how long it would take the Japanese to admit defeat, and what they might do in Indochina in the interim. And there were strong indications that the Japanese in Indochina were planning something other than surrender in the near future (Marr).

For their part, the Japanese had noticed the change in the temperament of the French. They'd learned that Mordant was the head of the planned resistance (Dommen). They noted that Decoux was noticeably less cooperative and more blatantly in the Free French camp. After the fall of the Vichy government in August 1944, Decoux had changed his letterhead to include the Provisional Government's *'République Française'* designation. In September 1944, with a major shortage of rice developing in Tonkin, Decoux had resisted when the Japanese authorities increasingly demanded rice for shipment to the Home Islands and other Japanese-held territories. Decoux said that if the Japanese wanted the rice, they would have to take it themselves (Gunn). In early February

1945 he noted in a radio address that France had been 'liberated', thus distancing himself from the still-extant collaborationist Vichy regime (Marr). Moreover, for some time Decoux had been taking steps to offer the Vietnamese better economic and socio-political conditions, to help 'immunise' them against the Japanese message of 'Asia For Asians' (Dommen). He had moved to improve opportunities and pay for native civil servants, and to provide better educational opportunities to their young people (Hammer). He also encouraged the teaching of local history (Dommen). He sought to reduce local corruption and provide better business opportunities for the commercial class (Marr; Dommen). In the parlance of a later time, he had been seeking to 'win the hearts and minds' of the ordinary people.

For most of the war, the Japanese found it cost-effective to allow the French to remain ostensibly in charge in Indochina, as long as they did what their Japanese overlords directed. In this way, the Japanese avoided the manpower and other costs of a full occupation, yet were able to extract rice and other resources for their own use. But by the beginning of 1945 the relative complaisance of the French was replaced not only by uncooperativeness, but also by apparent preparations for active resistance.

Aware of the changing situation, by January 1945 the Japanese in Indochina were increasingly concerned about several possibilities: an invasion from China by Chiang Kai-shek's GMD; amphibious landings by the Americans; or even that the Thais would switch sides just as Italy had done and thereby threaten the Japanese hold on Indochina from the west (Marr). General Tsuchihashi Yuichi, the military commander in Indochina, tried to prepare for these eventualities by commandeering more food, fuel and transport for his troops, as well as native labour to improve his defences, but the French authorities increasingly obstructed these efforts (Marr).

Adding to the Japanese concern were US Navy air raids in January on several Vietnamese targets. These included Saigon, the major Japanese naval port at Cam Rahn Bay, the airport at Tan Son Nhut and railway lines into Saigon (Dommen). The US Army also sent B-29s on 27 January and 7 February to bomb diverse targets throughout Indochina. Part of the purpose of these highly destructive raids was to divert the Japanese Navy's attention from the US landing on Luzon, 900 miles away in the Philippines (M. Thomas). The raids added to the belief by the Japanese, the French and the Indochinese that an invasion of Indochina by the Allies was imminent (Dommen).

Accordingly, on 1 February 1945 Tokyo approved Tsuchihashi's plan for a complete Japanese takeover of Indochina (Marr). French officials were to be detained and French and colonial troops and police would be disarmed and confined. Tokyo left the date for this *coup* to Tsuchihashi's discretion. Based on the anticipated arrival of the reinforcements he needed to accomplish this action, he scheduled the takeover for 9 March (Marr).

Noticing the French authorities' troop reshuffling over the previous several months, the Japanese had moved their own troops around to keep an eye on the French (Marr). In turn, Mordant now noted the Japanese (counter) moves. In early February, he also received word from French port officials that additional Japanese troops had begun arriving and would soon reach or exceed numerical parity with the Indochinese Army (Dommen).

Thus, Decoux and the other colonial leaders realised that doing nothing while waiting for the Japanese surrender, whenever it occurred, might not be a good option. That surrender was surely not in the distant future, but the damage the local Japanese might do in the short term to French interests in Indochina could be considerable. The Japanese appeared to be positioning their forces, and perhaps reinforcing them, to move against the French. How much mayhem might the Japanese unleash, in what could be a final eruption of violence? Given the Japanese history of abusing and slaughtering not only prisoners of war but also innocent civilians, Decoux and Mordant were worried about the welfare of the tens of thousands of French civilians in Indochina (Marr). In addition, it is possible that by mid-February Decoux and Mordant had learned about the specific Japanese takeover plans through Allied code-breaking (Dommen).

So, on 13 February 1945, the day the third Japanese city, Kokura, was destroyed by an atom bomb, Mordant ordered a series of attacks against the Japanese throughout Indochina. But instead of targeting Japanese garrison installations, he ordered the focus on supply depots, and rice storehouses in particular. He knew the Japanese had massive stockpiles of rice (Marr) even though Tonkin in the north was in the grip of a terrible famine. Japanese-mandated conversion of rice fields there to industrial crops like oil seed, cotton and jute had cut the 1944 spring harvest. Drought and pests further decreased that harvest. Then, typhoon flooding heavily damaged the autumn harvest (Gunn). The available rice was further depleted when the Japanese diverted some of it to make alcohol as a replacement motor fuel. Adding to the hardship for the people of Tonkin, January 1945 was unusually cold. The Japanese nonetheless continued taking large amounts of rice for their troops in Indochina and elsewhere and for shipment to the Home Islands (Dommen). As a result, by February 1945 hundreds of thousands of people in Tonkin had already died of hunger (Marr). Eight hundred miles to the south, Cochin China had escaped the crop failures and had surplus rice. But Japanese control of the single north–south railway and Allied air and naval activities along the coast prevented relief shipments by land or sea (Gunn).

As directed by Mordant with the acquiescence of his successor General Ayme, on 15 February French and colonial troops staged raids on the relatively lightly guarded Japanese supply bases and warehouses, especially in and near the port of Haiphong, near Hanoi. At some locations the French used heavy weapons they had kept hidden from the Japanese since 1940 (Fall). The munitions airdropped in by the British in late 1944 also helped. The French also struck at Quang Tri in Annam and Xuan Loc in Cochin China as well as at depots outside of Vientiane in Laos and Phnom Penh in Cambodia. The French and French colonial attackers, aided by the British and Free French agents infiltrated earlier, took losses, but in most cases gave better than they got. They came away with many tons of rice, and with weapons and ammunition that would further improve their limited firepower. What Japanese munitions they couldn't haul away, they destroyed, lest they later fall into the hands of the Viet Minh or the several other nationalist groups. Before the Japanese could organise counter-attacks, the French pulled back, usually deep into the forest (or in the case of Xuan Loc, into the vast Michelin rubber plantations east of Saigon). The rice liberated from the Japanese stockpiles soon began to find its way to villages throughout the hardest-hit famine areas.

The French raids were in a sense brilliant; in another, they were just common sense. General Mordant was not known as the most aggressive of generals, but the idea of a series of coordinated, high-profile raids against not-very-hard targets was within his comfort zone, especially given the expectation that a Japanese surrender was a matter of days or weeks away. Raiding food stores and distributing the rice to starving Tonkinese was an objective which the Vietnamese troops under his command could be counted on to execute valiantly. And the effect on public perception of the French 'liberation' of Japanese-hoarded rice could only be positive. This aspect fitted in well with Admiral Decoux's efforts to improve conditions for the Indochinese, to help smooth the return of full French control after the war. The French lightning campaign thus had an important 'hearts and minds' component. Decoux let it be known throughout Indochina that the French and their native colonial troops were fighting to relieve the Japanese oppression that was so wicked it was allowing people (fellow Asians at that) to starve to death.

Like the Russians' invasion of Manchuria in those same waning days of the war, the French action had not only military but also geopolitical significance. The Russians would claim that their action played an important role in pushing the Japanese to surrender; the French would make no such grandiose claim, but were pleased to point out that due to their timely actions, Indochina remained under French sovereignty throughout the entire war.

Despite some losses in the initial set of attacks, the French still had a numerical advantage over the Japanese in the key Tonkin area, and the munitions they had just appropriated would help overcome the Japanese firepower advantage. While the French repositioned themselves, Decoux broadcast news of his actions on 16 February. In accordance with his administration's alignment with de Gaulle's Provisional Government, it was his duty, he declared to his audience in Indochina and worldwide, to take up arms against the Japanese Empire.[1]

That Empire was preparing to strike back when its Emperor's message of surrender was broadcast on the 18th. Ostensibly frustrated by the order to cease hostilities, General Tsuchihashi may have been secretly relieved. Otherwise, he would have had to respond to the French attacks without the several additional divisions he needed for his planned March takeover. They had not yet all arrived (Kiyoko). Battling the fully energised French and colonial troops all over the countryside would have been costly. While he awaited surrender orders to trickle down the chain of command, Tsuchihashi shelved any plan to strike back at the French.

---

1. See 'What Really Happened' (9 Mar 1945).

Chapter 18

Indochina Restoration

The Viet Minh, Early 1945

In February 1945, the French were not the only group in Indochina working on an uprising. Indeed, insurrections and other conflicts would seem to be deeply embedded in the DNA of the various peoples of what Europeans called Indochina. One scholar has calculated that in the 900 years before the French arrived, there had been scarcely a year of complete peace throughout the region (Dommen). The Vietnamese have an especially rich, centuries-long history of rebellions against the government of the day (Karnow). For decades, the French had forcefully suppressed uprisings by various nationalist groups who agreed on little other than their opposition to the French (Duiker). The French colonial authorities nevertheless saw them all as tools of the communists (Dommen). The communist influence did indeed strengthen during the war. Before the war, a Vietnamese nationalist born with the name Nguyen Sinh Con formed the Indochinese Communist Party, the ICP, which was dominated by its Vietnamese members. Then during the war, using the alias Ho Chi Minh, he invited other nationalist groups, regardless of political ideology, to join an anti-French, anti-Japanese coalition. Ostensibly this organisation was focused more on liberation than on communist or any other political ideology. Officially titled the Viet Nam Doc Lap Dong Minh, Hoi, or League for the Independence of Vietnam, the 'Viet Minh', as it was commonly known, was still predominantly a communist organisation. It had close ties both to the Chinese Communist Party, and to Moscow (Cima).

During the Second World War, the Viet Minh conducted scattered guerrilla actions against the French, more so toward the end of the war, and more often close to the China border, but never in great force. The French periodically moved against them (and as mentioned, used them as an excuse for repositioning the troops of the French Indochina Army), but neither side did much damage to the other (Duiker). The Viet Minh focused more on organising at the province, district and village levels than on active resistance to either the French or Japanese (Fall). By 1944, the Viet Minh central committee programme called for a general uprising to seize independence whenever the Allies came, or the Japanese surrendered (Marr).

In late summer 1944 Viet Minh leader Ho Chi Minh returned to Tonkin from several years in China. His colleague Vo Nguyen Giap told him they were planning an imminent nationwide uprising. Ho advised them that the Viet Minh were not strong enough; only in Tonkin did they have some real strength, but that was also where the French troops were most concentrated. With Japanese concurrence, the French would be able to crush any rebellion, even in Tonkin. Even if the Viet Minh were successful against the French in some places, they certainly were not strong enough also to fight the Japanese. In October 1944, Ho wrote to his colleagues that he believed the opportunity for liberation would arise in about late 1945 or early 1946 (Marr). In anticipation of that opportunity, Giap began forming a 'professional' military force. In December 1944 he formed his first 34-man 'brigade' of the Vietnamese Liberation Army and with it he took two small French outposts, and their weapons, on 24 December. This small but symbolic action significantly aided Viet Minh recruitment among the rural population, as well as support among students and urban intellectuals (Marr). Several more small actions followed in January 1945 north of Hanoi (Duiker).

On 12 February, after they heard about the atomic bombings, the Viet Minh leadership met to determine what action to take. Ho Chi Minh was not present; he was once again in south China. (He had escorted a rescued American pilot shot down over Vietnam and was seeking to leverage that into American support for their anti-colonial struggle against the French.) Giap believed that the imminent end of the World War was the right time for the Viet Minh to make their move. Public declarations by the Americans such as in the Atlantic Charter, suggested that the US would support full liberation of colonial holdings such as Indochina. Perhaps Uncle Ho would soon return from China with such support lined up.

The Tonkin-born French scholar Paul Mus wrote that the Vietnamese have a well-established sense of the 'historical occasion, when all is temporarily possible and the world is plastic' (Marr). The Vietnamese even have an expression, *thoi co*, for such opportune moments. Some in the Viet Minh leadership believed that this was one of those moments. The Viet Minh, they argued, should issue an immediate ultimatum to all Japanese units to lay down their arms to the local Viet Minh cells. Had such a ploy been successful, the Vietnam Liberation Army would have netted considerable weaponry and munitions. Other members of the Central Committee realised, however, that the Viet Minh did not have the strength to follow through on any such ultimatum in any area where Japanese troops were stationed (Marr).

The Viet Minh by then did have a small 'liberated zone' in northern Tonkin along the China border (Dommen). The Viet Minh leadership reasoned that as soon as the Japanese surrendered, the Vietnam Liberation Army should be prepared to expand that base area and perhaps reach into the Red River Delta and even approach Hanoi. This would present the French and the other Allies with a *fait accompli* (Duiker). But this ambitious plan exceeded their means; the Viet Minh had very few weapons. They had sought arms from US OSS operatives, but by the beginning of 1945 had received almost nothing (Dommen). Some of their non-communist rival groups had secured weapons from the Chinese Nationalists, the GMD, benefactors who were not inclined to help the communist Viet Minh (Dommen).

Giap was planning for at least a limited northern offensive when the French began their warehouse-raiding campaign against the Japanese on 15 February. As noted, the French raids were successful, surprising the Japanese and liberating thousands of tons of much-needed rice. Giap and colleagues such as Pham Van Dong understood the significance of the actions by the French. They believed the French were obviously more interested in reasserting their position as colonial masters than in actually helping to defeat the Japanese Empire. The fact that the French were taking steps to save Vietnamese from starvation was, the Viet Minh leaders told themselves, just a cynical stratagem to diminish the people's resentment of colonial oppression.

The Viet Minh believed it was imperative that their movement now assert itself too. They would not defeat the French, but they could perhaps earn a voice in the post-war disposition of Indochina. They dispatched a message to this effect to Ho in China, but proceeded to act in his absence. On 16 February, a large portion of Giap's Vietnam Liberation Army, several hundred troops, attacked the small French outpost at Thai Nguyen, a site that had not been involved in the French action against the Japanese. Just as surprise had aided the French against the Japanese, so now too surprise assisted the Viet Minh against the French. They captured the outpost, but withdrew soon after, taking what weapons and materiel they could. The next day, a similar scenario unfolded at Cho Chu, albeit with less success against the more alert French garrison (Map 10).

The Viet Minh issued a call to all Vietnamese, urging them to rally to the cause of liberation. Thus in mid-February 1945 Indochina saw the Viet Minh attacking the hated French, the French attacking the hated Japanese, and the Japanese knowing they were defeated but not yet admitting it. All this was beyond most Vietnamese living outside of the major cities. In the north, the people's primary concern was not political but nutritional. By late February 1945, the complexion of the government was a lot less important to ordinary Vietnamese than how to feed their families. And the French were already distributing the rice they had liberated from Japanese storehouses in Tonkin, supplemented with shipments from the south.

It was hardly an original thought, but Admiral Decoux noted in a dispatch to Paris: *'La meilleure façon de les cœurs et les esprits des hommes est à travers leurs ventres'* ('The best way to men's hearts and minds is through their bellies').

And it was true. With French help, word spread that the Viet Minh were interfering with French efforts to get food to the people of Tonkin. After years of diminished crops due to Japanese demands, villagers felt that Frenchmen who brought food were a better prospect than Viet Minh cadres who talked about independence and did little else but levy their own taxes in rice.

French Restoration

On 18 February the Japanese Emperor broadcast to all his subjects his message to surrender. Soon thereafter, the Imperial General Staff informed MacArthur's headquarters that it would take some time for the surrender order to reach all Japanese troops, and perhaps still longer to persuade them of its authenticity. In Southeast Asia, it took six days for the Emperor's order to officially reach Marshal Terauchi's Southern Area Command headquarters in Saigon (Marr). Terauchi was to surrender his command, encompassing all of Southeast Asia, to Admiral Mountbatten. Further instructions from the Allies recognised the unique situation in Indochina, which the Japanese had technically never 'conquered' and where a French administration had remained in place throughout the war. Accordingly, the Allied order called for General Tsuchihashi's forces in Indochina to lay down their arms and place themselves under the direction of Admiral Decoux.

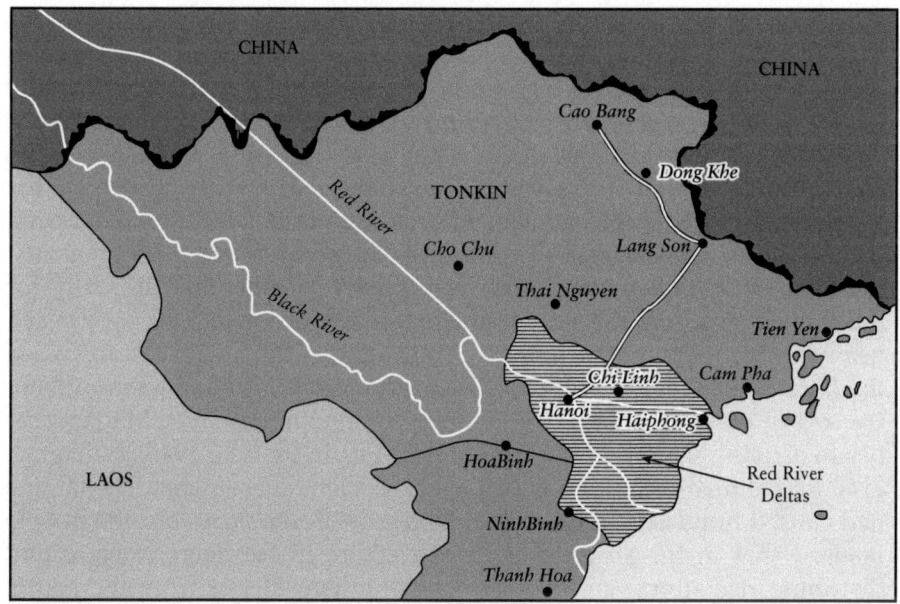

Map 10: Tonkin

Thus, after the formal surrender ceremony held by MacArthur in Tokyo Bay on 2 March, Governor General Decoux, representing the Government of France, and General Ayme, the official commander of French military forces in Indochina, received the surrender of General Tsuchihashi and all Japanese forces in Indochina on 9 March 1945. This was the very day Tsuchihashi had earlier planned to arrest the officers to whom he was now surrendering.[1]

The French swiftly upgraded the capabilities of their forces throughout Indochina, courtesy of the Japanese Imperial Army's stores of weapons, ammunition, vehicles, aircraft and fuel. Soon, Japanese-made trucks flying the French tricolour were speeding up the distribution of rice throughout the famished Red River Delta. At the least, the rice was welcome. In northern Tonkin, the Viet Minh's stronghold, well-armed French and colonial troops now escorted the rice trucks, in part to make sure the rice got to the people rather than the Viet Minh. The free flow of rice also served to encourage the free flow of information. There was perhaps no explicit coercion, but the amount of rice left with the villagers was in direct proportion to their helpfulness in identifying and locating

---

1. See 'What Really Happened' (2 Sep 1945 Indochina).

Viet Minh guerrillas and regulars. By the end of March 1945, hundreds of Viet Minh guerrillas and political cadres were in French custody. Vo Nguyen Giap's Vietnam Liberation Army had not been able to capture sufficient Japanese or French weapons to enable them to fight effectively. Anecdotal reports in later years from both Japanese and Vietnamese sources suggest that if the French had not acted swiftly to disarm the Japanese, a considerable number of Japanese units would have allowed their weapons and supplies to fall into Viet Minh hands. A well-armed Vietnam Liberation Army could have posed considerable difficulty for the French. Instead, the reinvigorated French Indochina Army exacted heavy casualties from the Viet Minh's troops on the few occasions when the latter were unable to escape a confrontation. Within two months of the end of the World War, fresh French colonial troops from Madagascar began arriving in Saigon (Logevall). As a result, by June 1945, the French exercised almost complete control over the Vietnamese states. The famine had caused hundreds of thousands of deaths in the north but the people realised that in the absence of French action in February to seize and distribute rice stores, and to expedite shipment of rice from the south, the death toll would have been much worse.[2]

There were grumblings and protests from a wide variety of nationalist groups. This included the communist Viet Minh as well as the non-communist Vietnam Nationalist Party and the Dai Viet group in the north (Marr), and the United National Front and the National Independence Party in the south (Dommen). Most resistance to the French, however, was peaceful.

The religious minorities, such as the Cao Dai and the Hoa Hao sects in Cochin China, sought autonomy from any central government, be it French, communist or imperial (Marr). Their 'militias' of several thousand troops each were irritating to the French administration, but not threatening (Cima). Likewise the well-armed Saigon-based Binh Xuyen crime syndicate (Karnow).

The French forces also maintained watch along the Tonkin-China border, to guard against possible movement of Chiang Kai-shek's GMD forces into Vietnam.[3] During the war, Chiang had been critical of French colonial rule (Marr). Although both the GMD and the French fought the Japanese, this was a case of the enemy of my enemy still being no friend of mine.

---

2. See 'What Really Happened' (Mar–Apr 1945).
3. See 'What Really Happened' (Sep 1945(2)).

> **A Coup Too Late**
>
> It is almost certainly apocryphal, but a 1960 document purporting to be reminiscences by Vo Nguyen Giap makes interesting reading. In it, he reflects on the post-war revelation that Japanese General Tsuchihashi would have taken over Indochina outright on 9 March 1945 if the war had not come to such an abrupt and unexpected end in February. 'The detestable Japanese would have done us a service by disarming and interning Mordant and all the rest of the French military. Then when the Japanese were defeated months later, our forces would have sought to bargain for, capture or steal from the Japanese the military supplies they had confiscated from the French. With French arms, and with the French disarmed and displaced, our movement would have been capable of filling the power vacuum, at least in Tonkin. If we had been able to declare establishment of a Democratic Republic of Vietnam, probably in Hanoi, in 1945, we could have prevented the French from so smoothly reasserting their control. How different the next decades would have been.'
> See 'What Really Happened' (9 Mar 1945 and 2 Sep 1945 Indochina).

In March 1945 de Gaulle's Provisional Government announced that imperial restructuring was central to post-war planning. Minister of Colonies Paul Giacobbi announced colonial reforms for Indochina. These included French citizenship, political rights and better employment opportunities for Indochinese within a proposed French Union (M. Thomas). The motivation for such reform efforts probably had multiple sources, including the diverse anti-French nationalist movements throughout Indochina, along with the anti-colonialist leanings of President Roosevelt, whose support France would clearly need for its post-war recovery.

The US position regarding French control of Indochina was not really clear. During the war, FDR had stated divergent positions about the future of French control of Indochina. In 1942, FDR led the French to believe that he endorsed retention of their colonial empire. But in early1943 he told the Joint Chiefs he had grave doubts about a restoration of France in Indochina. Then in March 1943 he discussed with British Foreign Secretary Anthony Eden a trusteeship for Indochina (Hess). During the Cairo Conference of November 1943, either from ignorance or disregard for the long and bitter pre-modern history of centuries of Chinese domination of Vietnam, Roosevelt discussed with

Chiang Kai-Shek the idea of China taking over control of Indochina to guide it toward independence (La Feber). But by the time of the Dumbarton Oaks planning conference in late 1944, the US was no longer so adamantly anti-colonialist. By January 1945, partly because of his waning confidence in Chiang's abilities, and partly because of the State Department's concern for the post-war importance of France as a strong ally, Roosevelt was prepared to defer any definitive decision on Indochina until after the war (La Feber). On New Year's Day he wrote to Secretary of State Stettinius, 'I still do not want to get mixed up in any Indochina decision. It is a matter for postwar' (Hess). The arrival of Harry Truman as President in April did not immediately clarify the US position in regard to Indochina.

After a meeting with de Gaulle, the new President wrote that 'he had received a satisfactory response from the General when he gave us his opinion that Indochina should receive its independence and that steps should be taken immediately with a view to arriving at that state' (Dommen). Putting aside the earlier concept of an Indochina Federation under a single French-appointed Governor General, De Gaulle instead intended that each of the Indochinese states would be given their independence (Dommen). For the sake of French national pride, however, this would happen through France's own volition (Miller and Wainstock).

There were those within the new French government in Paris who believed that the legal purges of Vichyists going on in the French homeland should extend to the colonial administration in Indochina. But de Gaulle took a pragmatic approach to prosecuting, and often pardoning, Vichyists in France, realising that a too drastic campaign would be counterproductive (Werth). His approach was similar in regard to Indochina. He believed that forcing Decoux and his colleagues out would be seen as punishing them for their lightning campaign that saved untold thousands of Vietnamese lives and knocked the potentially troublesome Viet Minh back into their jungle hideouts. De Gaulle retained Decoux as Governor General, and the administration stayed intact, thereby preserving a well-functioning colonial government that would have been crippled if the proposed purge of 'Vichyists' had been carried out. Such a weakened colonial government would have been less able to push back against the several nationalist movements gathering steam in the post-war period.

There was major flooding in Tonkin's Red River delta in August 1945 (Chandler et al.) Protecting against such periodic floods has long been a central responsibility of the government in Vietnam: maintaining the network of flood control dikes on the Red River is a constant requirement

(Marr). Because the Decoux colonial administration was intact and functioning, they had met this essential maintenance responsibility and minimised losses to the summer rice crop. If that had not been the case, that is, if the Japanese had displaced the French colonial government as they had planned, the resulting turmoil in local governance would have resulted in neglect of this essential infrastructure, and the consequent flooding damage to the rice crop would have risked renewed famine (Marr).

Bao Dai Restoration

In May 1945, the Provisional French government in Paris began discussions on the future of Vietnam with Emperor Bao Dai. During the war Bao Dai had two layers of government above him, that of the French colonial authorities and the Japanese occupiers. It was said that he reigned but not ruled, and did so from the imperial capital Hue, centrally located in Annam (Dommen). Now, the French had a position of some strength – they had finished the war still in control of the Vietnamese states and had scored a considerable success in their famine-busting campaign of the last days of the war in February. They had strongly re-established their ability to maintain law and order and suppress violent dissent. But the French also saw that a wide array of multiple groups was fanning nationalist sentiments and that there was little chance of returning to the *status quo ante*. The best chance for France to maintain a position in Indochina was to restructure the relationship with the several states and give their leaders, such as Bao Dai in Vietnam, the support they needed against the divisive factions they were facing.

Bao Dai also brought strengths to negotiations with the French. He represented the continuation of the Nguyen Dynasty dating from the early 1800s. So in the eyes of the Vietnamese, he was the legitimate ruler. Bao Dai had popular support from many of the increasingly vocal nationalist groups such as the Dai Viet, United National Front and VNQDD parties, along with the Cao Dai and Hoa Hao religious minorities. The latter two groups each maintained private militias of 10,000 or so troops (Dommen). However, Bao Dai's supporters did not include the communist-dominated Viet Minh (Dommen). Nonetheless, it was from this position of strength that Bao Dai addressed a message to France in the days right after the end of the war. In a firm but friendly tone he explained that the Vietnamese people would no longer accept foreign domination. 'I beg you to understand', he wrote, 'that the only means of safeguarding French interests and the spiritual influence of France in Indochina is to recognise the independence of Vietnam' (Dommen).

In December 1945, de Gaulle's government announced what came to be called the Élyseé Accords of 1945. This was an agreement with Bao Dai (and with the kings of Laos and Cambodia) to establish the Indochinese Federation under the French Union. Bao Dai would be head of state of all three Vietnamese states (Tonkin, Annam and Cochin China). While he had full control of domestic affairs, his foreign affairs were to be conducted in coordination with the other members of the Federation, and of the French Union, which was led by the President of France. The agreement included a provision under which any of the Federated states could opt out after ten years and become wholly independent.[4]

Truman and his State Department were pleased that de Gaulle had followed through on his representations to the President by setting up a transitional trusteeship without calling it so. The US was among the first countries to recognise the State of Vietnam. Bao Dai won the support of almost all of the nationalist groups, even though most of them would have preferred immediate full independence. The Viet Minh, from their jungle hideaways, denounced the agreement as perpetuating colonialism. But in a land that had endured 900 years of Chinese domination, it was hard for the Communists to paint a further ten years of French oversight as an intolerable burden, especially after decisive French action had saved many thousands of Tonkinese from starvation early in the year, and continued French administration had helped prevent potentially devastating Red River floods in August. Soon after announcement of the Accords, Bao Dai told his nation that he would use the next ten years to rebuild the strength of the nation, and develop efficient government capabilities with an eye toward exiting the Federation and the French Union by 1 January 1956.

The Emperor formed a government that included representatives of many of the nationalist groups, including the VNQDD and Dai Viet, and also the Cao Dai and Hoa Hao (Dommen). He offered the post of Prime Minister to Ngo Dinh Diem, a prominent former provincial governor. Diem had served briefly in a pre-war Bao Dai government, but had resigned over his concerns about French control of the government (Fall). Now again the same concerns prompted Diem to decline to serve in an administration which he still considered colonialist. In subsequent years, Diem, an ardent Catholic, attempted to form an anti-communist, anti-French Third Force in

---

4. See 'What Really Happened' (Mar 1949).

politics (Miller 2013). It never gained any real traction and Diem remained a bitter, divisive, but minor figure in Vietnamese politics.[5]

Instead of Diem, Bao Dai tapped Nguyen Ton Hoan, another Catholic anti-communist nationalist. A leader of the Dai Viet party, Hoan wanted an end to French involvement in Vietnam, but he believed that working toward that end with Bao Dai would be in Vietnam's best interest (E. Miller 2004). Hoan reasoned that a broad-based government representing most of Vietnam's diverse geographic, religious and political interests could keep enough pressure on the French to ensure they kept their promise of full independence for a united Vietnam.

As for Bao Dai himself, he'd been frustrated by his pre-war years of being the monarch of a French protectorate, and then enduring the war with the Japanese as an additional layer of control atop the French. Some observers said that the French-educated Emperor seemed more interested in the pleasures of the flesh than in the labours of leadership. Yet he was astute and skilful, with a keen understanding both of his countrymen and of the French (Dommen). Now in the post-war era he had the opportunity to help his nation regain its pride and position in the region, and he proved the observers wrong about his dilettante preferences. He devoted most of the rest of his life to his country. He wanted Vietnam to be free of French control, although he did not categorically despise the French as some of his compatriots did. He looked forward to good relations with France as a trading partner, and as a power that could help protect his country from Vietnam's ancient and major enemy to the north. Once the internal struggle in China was resolved, that domineering nation might once again menace Vietnamese freedom. Therefore retaining France as an ally would be in Vietnam's interest. A lesser security concern to Bao Dai was the annoying Viet Minh group in the borderlands near China.

---

5. See 'What Really Happened' (Oct 1955).

Chapter 19

Indochina Independence

Campaign against the Viet Minh

As 1946 began, the Viet Minh were weak, having lost many of their guerrilla fighters and cadres in battles with the French at the end of the war. The survivors were ill-equipped; they had a few weapons seized in skirmishes with the French, and a few more acquired from defeated Japanese troops. Their fellow Marxists to the north could not provide much help: in the midst of the renewed Chinese Civil War, Mao Zedong's communist movement was in no position to provide materiel assistance to their Vietnamese comrades. The Viet Minh were also ill-fed. The farmers of Tonkin were not inclined to feed the communists who had tried to interfere with the French troops' food relief efforts at the end of the war. And the communists had only limited means with which to coerce supplies from the villagers. Once word spread that Bao Dai had secured Vietnam's autonomy as a transition to full independence in just a few years, the Viet Minh's anti-colonial message lost most of its relevance. Still, Ho Chi Minh, Vo Nguyen Giap and the other Viet Minh leaders were committed to the communist cause, even though they had tried to camouflage it by calling their movement a coalition of nationalist groups.

The Viet Minh garnered some support from the Tho and other minority tribes in the northern mountains, exploiting those groups' resentment of their treatment by Imperial Vietnamese and French administrations alike. Pushed out of the lowlands in times past, the minorities typically relied on slash and burn agriculture, which is much less productive than rice farming in the delta. The lowland people then denigrated the hill people for being less hard working (Filippelli). (There are hundreds of Native American tribes who would have no trouble understanding that dynamic.) But neither did the hill peoples typically share the Viet

Minh's somewhat esoteric ideological fervour, so the relationship was not much more than tolerance.

By late 1946 Bao Dai's government had accepted the transfer of many units of the French Indochina Army into a new Vietnam National Army (VNA). They were receiving updated weapons and equipment, much of it via US Lend-Lease supplies to the French (Marr). In the south, several thousand troops from each of the minority sects, the Cao Dai and Hoa Hao, were aligned with the VNA. These combined forces were reasonably effective in protecting most of the country from the occasional raids, and attempted raids, staged by the Viet Minh.

In January 1947 General Jean-Étienne Valluy, the commander of the French Indochina Army, launched Operation Lea. This was a large-scale operation involving French paratroopers, colonial land and riverine forces, and French fighter/bomber aircraft. Focused on several villages northwest of Hanoi, the operation aimed to decapitate the Viet Minh movement (Fall). Valluy had intelligence that Ho Chi Minh, Vo Nguyen Giap, Pham Van Dong and all or most of the rest of the Viet Minh leadership were in the village of Bac Kan (Filippelli). Valluy's agent was correct, and the paratroopers stormed into the village just as the surprised Communists were hastily packing maps and papers. The paratroopers killed or captured almost the entire leadership of the Viet Minh. A few escaped the village, only to be caught by the motorised infantry advancing from Cao Bang. It was a devastating blow.[1]

In later years, the exiled Vo Nguyen Giap reflected that perhaps the greatest shortcoming of his career was his inability in 1946 to recruit a reliable agent to work in the French headquarters.

> If we had succeeded in placing an agent there, as we had been trying to do, we would likely have gotten at least some warning of that raid, and we could have evaded the French that day. But by that time, our movement was weak, with so little support from the people. We had very little success in enlisting personnel from the government, or in infiltrating our own comrades into key positions. If we had placed an agent who could have warned us, perhaps our glorious struggle would have been able to continue past that critical moment. What effect our escaping Bac Kan would have had on the subsequent history of Vietnam we cannot know.

---

1. See 'What Really Happened' (Oct 1947).

As Giap said, we cannot know. If the Viet Minh leadership had escaped Valluy's Operation Lea, it is possible that the communist movement could have held together in the Vietnam-China borderlands for some time, perhaps until Mao Zedong's Chinese communists were able to provide support, training and equipment to aid their Vietnamese comrades. Such help from the Chinese could have made a huge, even decisive difference in Ho Chi Minh's struggle against the French (Gaddis 1997). But after Operation Lea in early 1947, the Vietnamese Communist Party was decapitated, and the remnants could not sustain themselves as a fighting force for the five years before the Chinese communists had won their civil war and would have been able to provide aid to the Vietnamese comrades. In 1949, several of the surviving senior Viet Minh leaders travelled to China to meet with the Chinese communists to seek their assistance (Westad). Mao Zedong was favourably inclined to help his fellow revolutionaries. But, he told Party Chairman Lê Duân, he was still too embroiled in fighting the Guomindang for control of China to spare the resources, nor could he have gotten any important amounts of materiel or personnel to Vietnam, since the Guomindang still controlled Yunnan Province on Vietnam's border.[2]

Finished as a fighting force after Operation Lea and with no help available from the Chinese comrades, the surviving true believers of the Vietnam Communist Party continued trying to spread their Marxist message throughout the countryside, primarily in the north. They drew few supporters. Communist attempts to coerce isolated villages into supporting them drew swift and effective response from the VNA and police forces of the Bao Dai government. In the south, where the communists had even less of a presence, they also clashed with the Cao Dai sect's troops (Dommen). General Valluy, in coordination with French High Commissioner Emile Bollaert and the Bao Dai government, primarily used his French soldiers to train and assist Bao Dai's troops. In this way, ordinary Vietnamese seldom saw French soldiers, and villagers came to see the Vietnamese troops as their protection against communist intimidation and thievery.

The government efficiently maintained the dikes in the flood-prone Red River delta – one of the most densely populated rural areas in the world. They improved the north–south railway and road network, to ensure that rice could be redistributed between the Red River and the Mekong River farming regions as needed, lessening the famine effects

---

2. See 'What Really Happened' (May 1950).

of flooding in one or the other region. (Although both the Red and the Mekong River deltas produce abundant rice with the help of seasonal monsoons, the rains that fall in the north are part of a different climate regime than those that fall 800 miles to the south. This means that flood or drought conditions in one region are not necessarily accompanied by the same in the other region [United Nations Environment Program].)

With the help of the French Navy, the government crushed the pirate operations that interfered with coastal shipping. Especially in the north the government also spread the message that the Viet Minh were communists more than they were nationalists. These Viet Minh, the government messaging went, had strong affiliation with, and were being supplied and aided by the world communist movement, and especially the Chinese communists. So helping the Viet Minh was therefore helping Vietnam's ancient enemy, China. By contrast, working with the government was advancing the imminent prospect of full independence from France, with the further prospect that France would thereafter remain an ally to help keep the odious Chinese out.

Independent Indochina

In 1949, after a Constitutional Assembly, Vietnam became a constitutional monarchy. In elections in 1950, Ngo Dinh Diem challenged Nguyen Ton Hoan for the premiership. Diem campaigned for immediate exit from the French Union. In addition to his own longstanding dislike of the French, his position was based on the decolonisation sentiments then welling up in other French-controlled lands such as Madagascar, Morocco and Algeria (Metz 1993, 1994). In Madagascar, for example, the French had deployed 30,000 colonial troops to suppress the bloody Malagasy Uprising of 1947–8 (Leymaire). Diem's 'Independence Now' message gained some support, but his perceived religious divisiveness as an intolerant Catholic in a majority Buddhist nation, lost him more. Hoan won the election, and promptly began discussions with the French about exiting the French Union sooner than the original January 1956 target.

As the 1950s began, Mao Zedong's Chinese communists were gaining the decided advantage over Chiang Kai-shek's Nationalists. So the possibility of a communist giant as Vietnam's neighbour weighed on the minds of Vietnamese, French and other Western leaders. With an eye on developments in China, the French government of Premier George Bidault and the Vietnamese government under Nguyen Ton Hoan agreed in June 1951 that France would recognise Vietnam as a fully independent nation. In 'separate' accords, the two nations offered each other preferential trade and investment privileges, and Vietnam agreed to the maintenance of French military forces at bases at Haiphong and Cam Ranh Bay. Unpublicised at the time was the agreement by the Truman administration to increase its aid to France under the Marshall Plan by essentially the same amount as the cost to France of maintaining naval and other military units at bases in Vietnam. The US sought to set up containment in the event that a communist China warranted it.

As an independent nation with substantial resources Vietnam was able to develop its economy in ways that served its interests, rather than those of its former colonial master. The decade of the 1950s saw the development of coal mines in the north, hydropower in the central highlands, and offshore oil and gas in the south. From the mountains came iron, antimony, aluminium, chromium, gold, phosphate, tin and zinc. The Red River delta in the north and the Mekong delta in the south produced rice enough for export, while other farmlands produced exportable tea, coffee, cotton and rubber, and the dense forests produced timber. The long coastline and many lakes and rivers yielded abundant fish (Cima).

Along with economic development there was military preparedness. Into the early 1950s the remnant communist Viet Minh insurgency was an irritant in the region bordering China. With US aid (channelled through France), and French military advisors, the Vietnam National Army and police forces established a strong and effective presence in those borderlands, largely preventing the Viet Minh holdouts from receiving much shelter in China or assistance from their Chinese comrades. In the 1955 elections, members of the Communist Party of Vietnam emerged from their jungle hideouts and ran for local offices. Their electoral support was negligible and remained so into modern times.

By the mid-1950s the Vietnam military were less concerned about insurgents than about direct aggression by its northern neighbour. This was not so much based on Mao Zedong's ideology as on millennia of persistent enmity between the Vietnamese and the Chinese. The last time the Chinese had entered Vietnam, it took 900 years to get them out (Karnow). The Vietnamese had gotten free of the Chinese in the tenth century, but for a nation with 5,000 years of civilised history, the centuries-old memory was still strong. Now the Vietnamese had the backing of Western powers to help keep the Chinese out. This was formalised in the Southeast Asian Treaty Organization. SEATO, which included France, Great Britain and the US, was a counterpart to NATO, seeking to contain expansion of communism in East Asia (Department of State 1954).[3] The inclusion in the organisation of not only Vietnam, but also of Laos, Cambodia and Thailand helped present a united front against expansionism on China's southern flank. It also effectively discouraged disputes among the treaty members, each of whom had been involved in at least one border quarrel in recent decades. Thus, Vietnam was peaceful through the 1950s and beyond.

---

3. See 'What Really Happened' (8 Sep 1954).

PART III: POST-WAR SCIENCE

Chapter 20

Nuclear Research after the War

The Manhattan Project

The scientists and engineers at Los Alamos, the bomb development facility in New Mexico, had mixed feelings at the end of the Second World War. There was pride in their role in helping end the war, but regret too. The atom bombs caused appalling death and destruction. It was even worse that neither two, nor four such weapons prompted the enemies' surrender. Six weapons fell before the Nazis and the Japanese capitulated. Even then, it could be argued, the ongoing conventional military pressure from the Russians and the Western Allies was critical to forcing the surrenders. While it was clear that the atomic bomb was a much more awful weapon than anything before, 80 years previously the same had been true of Alfred Nobel's invention, dynamite. Then, Nobel had hoped that his new explosive would make war too terrible to fight (Rhodes 1995). Now, two World Wars and many smaller ones later, a few of the Los Alamos team thought that *their* invention really would put an end to warfare. Others were more pragmatic about the prospects for future wars, but decided that they had contributed enough to the science of death and returned to civilian life (Rhodes 1995).

A few of the Manhattan Project team turned their attention to the further development of nuclear weapons. They pursued two tracks: one was advancing the design of the atomic bomb. Perhaps the bomb would never be used again; but perhaps also the United States would need to possess a large-enough arsenal of such weapons either to deter the next war, or to win it. The six bombs that had been used were essentially prototypes produced at huge expense by the Manhattan Project complex. Reducing the cost of atomic weapons was one objective. Reducing their size and weight was another, so that if future bombs ever had to be used, they would not require the specially-modified 'Silverplate' B-29s, which were stripped of their armament and had their bomb bays reconfigured to be able to carry one bomb apiece. Most of Los Alamos' attention in the immediate post-war period was on this next generation of nuclear fission bombs (Rhodes 1995).

The other track was an effort to develop yet another new type of weapon. Early in the war, physicist Edward Teller became convinced that large amounts of energy could be released not just by fission (splitting of the large nuclei of heavy elements), but also by fusing together the small nuclei of light elements, such as hydrogen. His calculations suggested that fusion of hydrogen nuclei would release many times more energy than the splitting of uranium or plutonium nuclei. To the annoyance of others at Los Alamos, Teller spent much of his time during the war working on how to force hydrogen nuclei together and create 'the Super' as he called it (Rhodes 1995). By late 1944, Vannevar Bush and James Conant, of the Office of Scientific Research and Development, wrote to the Secretary of War about the likelihood of such a weapon being developed (Bush and Conant).

Oppenheimer's successor as Director at Los Alamos was Norris Bradbury. With the post-war realisation that atomic weapons might not be terrible enough to fulfil Alfred Nobel's vision of a war-preventing weapon, Bradbury reviewed Teller's idea. He then allocated research resources to efforts to develop 'the Super'. Not everyone in the nuclear community supported the move. Manhattan Project scientists such as I.I. Rabi, Fermi and Oppenheimer himself, did not believe they should develop 'the Super' (Sheehan). Others within the community noted that six bombs (apparently) had to be used before the Axis finally gave up. This meant, they said, that terrible as they may be, atomic bombs were at best war-enders, not war preventers. But 'the Super', the hydrogen bomb, now that would surely deter wars.

The argument prompts speculation: might fewer bombs have had a more potent long-term effect? That is, if just one or perhaps two bombs had been used and had (apparently) ended the Second World War, would such weapons then have been held in higher dreadful regard? Would they have been seen therefore as such extraordinary weapons that no further development was warranted?

Perhaps. But probably not.

Klaus Fuchs was among the Los Alamos researchers who did work on the hydrogen-fusion bomb (or H-bomb) and he made important contributions to its design. With John von Neumann, he filed a patent for that work (J. Bernstein 2010). So the US benefited from Fuchs' work. But the Soviets benefited even more: Fuchs later passed to them much of the H-bomb design data (Goncharov).

Soviet Nuclear Efforts

With the end of the war, and the advent of the atomic bomb, Stalin directed Igor Kurchatov and his team of academicians to accelerate the bomb research project they had started during the war. He wanted a Soviet bomb as quickly as possible (Rhodes 1995). As noted earlier, the Western Allies' use of the atom bomb made a huge impression on Stalin, as perhaps was intended. He understood it both as a weapon of war, and as a geopolitical tool. He saw that the US, UK and Canada had collaborated closely in its development, but had not invited their Soviet ally to participate. He believed this was evidence of their ill-will and their intent to practise atomic blackmail (Gaddis 1997).

In 1942 Stalin had set up a nuclear effort after Soviet researcher Georgy Flyorov informed him that the several scientific journals which had been consistently reporting on nuclear work by German, British and American scientists had suddenly gone silent on the subject (Kean). Flyorov asserted that this absence of information was evidence that the other nations were pressing ahead on secret nuclear weapons work. The result was Igor Kurchatov's 'Laboratory #2' of the USSR Academy of Sciences, a pale analogue to the Manhattan Project. Its work was hampered by a lack of uranium ore within the Soviet Union. That, and the fact that the Soviet Union was in an all-out life and death contest with the Nazis.

Later, Laboratory #2 benefited from information provided by Soviet spies within the Manhattan Project. However, Lavrenti Beria, head of the Soviet security apparatus the NKVD, suspected that the material from Los Alamos was disinformation (Rhodes 1995). So he sought to have Kurchatov and his colleagues independently develop a bomb with minimal reliance on the Manhattan Project information. But after the bombings of Dresden and Leipzig (in the Soviet zone of occupation) Soviet scientists had the opportunity to inspect the bombed areas and take samples of air, soil, and water. Analysis of the Dresden samples for fission products confirmed what would be expected from the use of a uranium device as described in the stolen design data. Data from Leipzig's ruins similarly confirmed a plutonium device. These findings helped persuade Beria to allow Laboratory #2 to make more use of the stolen data without requiring so much time-consuming re-invention efforts.

The Red Army's capture of the Germans' uranium-processing plant and the stores of uranium ore at Oranienburg also helped. As noted earlier, General Groves had tried unsuccessfully to get this facility destroyed before the Soviets could get to it. The advanced chemical engineering and metallurgy capabilities of the Oranienburg plant saved the Soviets a significant amount of time and effort in their bomb development effort.[1]

While the Pentagon knew about the Soviets having the Oranienburg equipment and stores of uranium ore, they did not know the extent of the espionage penetration of the Manhattan Project. The US estimated that it would be at least the early 1950s before the Soviets had the bomb (Rhodes 1995). Nonetheless, the Atomic Energy Commission (the civilian agency established to conduct nuclear weapons research and development) collaborated with the Air Force in 1947 to establish a Long Range Detection Program. This was to check the atmosphere for radioactive materials that would indicate Soviet nuclear testing (Battle). So it was a surprise, in the summer of 1948, when Air Force planes detected residue from the Soviets' first atomic explosion.[2]

Harry Truman released the news promptly:

> I believe the American people, to the fullest extent consistent with national security, are entitled to be informed of all developments in the field of atomic energy. That is my reason for making public the following information. We have evidence that within recent weeks an atomic explosion occurred in the USSR. Ever since atomic energy was first released by man, the eventual development of this new force by other nations was to be expected. This probability has always been taken into account by us (Truman 1950).

Mindful of his underdog status in his re-election race, Truman went on to note that the US atomic programme had continued prudently in the post-war years. He assured the American people that the nation had prepared for the loss of its atomic monopoly by producing sufficient nuclear weapons to ensure that no nation could threaten American security. He disclosed that he had 'previously' directed the Atomic Energy Commission 'to continue its work on all forms of atomic weapons, including the so-called hydrogen or super-bomb' (Truman 1950).

Critics suggested that Truman's 'previous' direction to the AEC to proceed with the H-bomb might have been the day before his public announcement in the summer of 1948, but that was not important.[3] What mattered was that the US was and would remain far ahead of the

---

1. See 'What Really Happened' (15 Mar 1945).
2. See 'What Really Happened' (29 Aug 1949).
3. See 'What Really Happened' (Jan 1950).

Soviets in nuclear weapons. In November, Truman won re-election in arguably the greatest election upset in US presidential history – up to that time (I. Ross). It is unknowable what role his strong stance on nuclear weaponry played in his election.

The US intelligence community believed that the Soviets would immediately press on with developing an H-bomb, and not just because the US was doing so. The consensus was that the USSR had very little uranium, so they couldn't build an extensive stockpile of atomic weapons. Therefore they would seek to develop hydrogen-fusion weapons, which would deliver far more explosive power using far less uranium (Rhodes 1995). Indeed, 'hydrogen' devices still need uranium, or plutonium made from it, because the explosive release of energy that results from fusing hydrogen nuclei only occurs at the sun-like temperatures created by a fission explosion. That is, an H-bomb requires an A-bomb to set it off.

Only years later did the US learn that the Soviets had obtained uranium not only from the Joachimstal mine in Czechoslovakia but also from East Germany, and from newly-discovered deposits in Siberia (Rhodes 1995). So it wasn't a shortage of uranium that prompted Soviet H-bomb development, but mostly a perceived need to match US capabilities.

In those early post-war years, the Soviets scrambled to develop not only their own nuclear weapons but also a way to deliver them. For its part, the US obviously already had aircraft capable of delivering atomic weapons to distant targets: the B-29s that dropped the bombs on Japan flew 1,500 miles from Tinian, about the same distance that UK-based US bombers would have to fly to reach Moscow. The Soviets had a geographic disadvantage. None of their allies at the time could provide airfields from which Soviet bombers could reach the US homeland, even if the Soviets had long-range bombers comparable to the B-29. And in the immediate post-war years, the Russians had no such aircraft.

Therefore in addition to their atomic research and development programme, the Soviets also undertook urgent programmes to develop delivery systems. One such system was to be a long-range bomber

---

ORDERS OF MAGNITUDE

A hydrogen bomb is not just a somewhat bigger nuclear bomb, like a 155mm artillery round is somewhat bigger than a 105mm round. An atomic bomb is already thousands of times more powerful than a conventional bomb, and a hydrogen bomb is a thousand times larger than that.

very much like the B-29. In fact, it would be a close copy of several US B-29s that were forced to land in Soviet territory during the Second World War. (On bombing runs from China to Japan, these aircraft had experienced equipment problems that precluded them making it back to friendly territory in China, so they landed near Vladivostok [Gordon and Rigmant]). By 1949 the USSR had a fleet of these reverse-engineered bombers dubbed the Tu-4, but even with them, bombing the US homeland would be a one-way mission for the Soviet crews. The new Russian bomber did not have enough range to get back home, and the Soviets had not yet mastered the aerial refuelling capabilities already demonstrated by the USAF (Johansen).

The other way to drop an atom bomb onto America would be by ballistic missile. In 1945, no nation had a missile with intercontinental reach, let alone one that could heft an atomic warhead. But the Soviets were closer than anyone else.

Chapter 21

The Rocket Men – Germans and Russians

As early as 1940 Wernher von Braun and his team at Peenemünde did preliminary design work for a two-stage rocket much bigger than their V-2. This 'A9/A10' missile was to be powered by multiple engines, each far larger and more powerful than the V-2's. As envisioned, this missile could have reached New York from France. It would have carried a 2,000lb warhead (J. Neufeld). There were also early ideas for adding a third stage ('A11') to create a satellite launch vehicle, and a fourth stage ('A12') to launch a manned orbiter (Wade). Had work continued on the A9/A10, this first intercontinental missile might have flown in 1946 (Wade). But in 1943 the German Army ordered the rocket team to focus exclusively on the V-2 (Wade). Late in 1944, work on the A9/A10 resumed under the code name 'Projekt Amerika', but no hardware work occurred before the end of the war (Wade).

Now von Braun and most of his team, along with several tons of their documents and design drawings, were in Soviet hands. The Soviets knew that the Germans' work was many years ahead of their own, and was perhaps ahead of the Americans'. The first Russian engineers to inspect the V-2 called it 'that which cannot be' (Ward). This from accomplished rocketeers themselves: before the war, the Russians had done pioneering work in rocketry, especially under Konstantin Tsiolkovsky in the early 1900s and in the 1930s under Sergei Korolev, Valentin Glushko and others. That work was severely impaired by Stalin's purges, during which both Korolev and Glushko were imprisoned (Heppenheimer). During the war, Korolev, still technically a prisoner, worked on rocket boosters to help heavily-laden aircraft take off. But as the Nazis' rocketry accomplishments became known late in the war, the Party determined that Korolev's advanced expertise was needed elsewhere (Heppenheimer). So it was that as the Red Army advanced toward Berlin and Peenemünde in early

1945, newly-released and commissioned Lieutenant Colonel Korolev and a team of colleagues followed close behind (Heppenheimer). They were armed with lists of names and materials to look for. High on that list was Wernher von Braun, whom the NKVD had been keeping their eye on since before the war (Brzezinski).

Very soon after the Red Army captured von Braun's convoy near Peenemünde on 17 February 1945, Sergei Korolev was on the scene.[1] He quickly realised the convoy's value and explained it to the commander of the division that had seized it. He stressed the importance of safeguarding all the personnel, the equipment and the documents in the convoy. He emphasised that even the SS officer who was captured at the head of the convoy, Sturmbannführer von Braun, was to be treated decently, at least for a while.

Korolev soon took from von Braun the orders he had been carrying. The orders showed the convoy was heading for Nordhausen, about 220 miles to the south-west.

'Why?' Korolev asked von Braun, 'What's in Nordhausen?'

From the first moments he found himself in Red Army hands, von Braun knew he had to decide how he was going to deal with his capture by the Russians instead of by the Americans, as he would have preferred. As a loyal German would he attempt to stonewall against these despised Slavs? What would that mean for his own future and that of the team he had led through the war? Would he have a future measured in more than hours? He was a Prussian of noble birth, and manifestly a Nazi field-grade officer, neither fact likely to be valued by the communists. But perhaps his rocket expertise would be. Of course, the Reds had just captured tons of equipment along with plans and blueprints that embodied a great deal of the expertise he and his colleagues had to offer. So it was entirely conceivable that the Soviets would feel they could dispense with him and his team. The documents wouldn't need to be fed.

Von Braun expected that he would be asked about his destination, and he knew full well what was at Nordhausen. His own brother Magnus was a production manager at the underground Mittelwerk complex (Ordway and Mitchell). But the real reason he had forged those orders 'directing' him to Nordhausen, was to try to get closer to the advancing Americans and thereby fall into their hands when the end came. If he tried to deny any knowledge of what was there or why he

---

1. See 'What Really Happened' (Late Feb 1945 and 2 May 1945).

had been ordered there, that lie would very likely be found out. There were too many references to Mittelwerk in the captured documents, and in the heads of the captured team. If he attempted some obstruction, and thereby prevented the Red Army from reaching Mittelwerk before the Americans, he would have little chance of surviving the next few months. But by telling his captors about Mittelwerk, and the technological treasures it contained – V-1s, jet engines, V-2s and other items – he would be demonstrating that there was value to his new masters in keeping him alive and cooperative. He would also be foreclosing his brother's chance to fall into the hands of the Americans. Still, he could perhaps collaborate his way to continuing the pursuit of his passion for exploring space, even if that meant working for another government developing ways to drop explosives deep inside their enemy's homeland. We do not know how long it took him to make up his mind, but by the time Korolev asked von Braun about Nordhausen, he was ready to reveal that it was the V-2 factory.

Years later, a rocketeer who worked closely with von Braun commented 'He doesn't care what flag he fights for' (Ward). Also years later, the wife of Helmut Grottrup, a senior member of the rocket team, wrote that von Braun had previously made a bargain with the Nazis. Von Braun would develop weapons for them, which would enable him to be involved in rocketry. So Grottrup's wife was not surprised that he would 'make another bargain rather than go off to the *gulag*. Beyond family and friends, rocketry and spaceflight were all that really mattered to him' (M. Neufeld 2007). It was also said of von Braun, 'He aimed for the stars, but he more often hit London' (M. Neufeld 2007).

On 18 February Korolev passed to his superiors von Braun's information about Nordhausen, along with an urgent plea for Red Army troops to get there as quickly as possible. The German surrender the next day made it possible for Red Army troops to dash westward toward Nordhausen, near the previously agreed-to western border of the Soviet zone of occupation. Troops of Marshal Konev's First Ukrainian Front entered Nordhausen on 26 February and quickly focused on securing the Mittelwerk complex. As a lower priority, they provided some food and care for the slave labourers. If any Soviets noticed the markings on some of the food items that the 'errant battalion' of the US 104th Division had left for the labourers three days prior, they took no note of it.

Korolev and Stalin were pleased that most of the German rocket expertise and material was in Soviet hands. Over the next several weeks, Korolev and other Soviet rocket engineers did a preliminary review of the captured V-2s and their documentation. They also interviewed von

Braun and the other Germans. Among themselves the Soviets saw a series of improvements they could make in the V-2 to enhance its range and improve its accuracy (Cadbury). Korolev was disconcerted during one of his interviews with von Braun when the German volunteered almost all of the V-2 improvements the Soviets had come up with. Von Braun said that getting the V-2 into production under enormous pressure from Berlin had precluded incorporating many design improvements that he and his team had themselves identified early on. 'To us, the V-2 was really just an initial, rudimentary design', von Braun said, 'We had already planned to avoid these shortcomings of the V-2, as you'll see from our advanced designs' (Cadbury).

'Advanced designs?' asked Korolev.

Von Braun then described the A9/A10 intercontinental missile, and the A11 satellite launcher and the A12 manned orbiter. Von Braun reasoned that the rough designs for these advanced vehicles were in the piles of documents that the Soviets would eventually plough through anyway. So he was earning goodwill, he hoped, by giving the Soviets something now that they'd later find anyway. 'Early in the war we had hoped to fly the Projekt Amerika rocket by 1946', said von Braun, calculating that the Soviets would see the value that he and his team could continue to provide beyond the data already committed to paper. 'We did not have a schedule for the satellite launcher,' von Braun told Korolev, 'but of course we knew that putting a satellite into orbit would open important new military, as well as scientific opportunities.'

As the German rocket man said that last piece, he sensed that for Korolev, just as for himself, the prime objective was getting into and exploring space, not using it for destructive purposes. Von Braun surmised that he and Korolev both knew the only way they would get to pursue their cosmic dreams was by building truly down-to-earth weapons for their employers. Perhaps von Braun further reasoned that if he and his team developed sufficiently impressive weapons, those weapons were not likely to be used. He believed that Robert Goddard in the US had been leading a major wartime rocket programme (M. Neufeld 1996). This may have prompted von Braun to believe that there would be an immediate post-war rocket race between the Soviets and the Americans. If so, he could have reasoned, making sure that his new Soviet employers could match the Americans missile for missile could be the best way for him to help prevent another war.

Then again, it is also possible that these geopolitical considerations were of no concern to von Braun. It is possible that his most urgent concern was survival for himself and his colleagues, preferably not in a prison camp, and ideally being able to resume their rocketry work. Sensing a kindred dreamer spirit in Korolev, von Braun may have suppressed any qualms he had about the short-term military purposes to which his space exploration vehicles might be put.

For his part, Sergei Korolev was initially not favourably inclined toward the German rocket men. Despite his near-death experience in the *gulag*, Korolev was a loyal Soviet citizen. Accordingly, he detested Germans and the devastation they had wrought on his nation. So he tried hard to find shortcomings in their rocketry, and improvements he could make. For years afterward, his first reaction to any new design or suggestion from most of the Germans was to be highly sceptical of it, if not to reject it out of hand. But he understood Soviet policy well enough to know that making full use of the information and expertise of others was an important and expected part of Soviet technological development. So he could not let his Russian pride prevent him from exploiting the German treasure at hand. He had to overcome his gut reaction to dislike the Germans and their ideas. As long as he stayed clearly in charge, Korolev would put von Braun and the others to best use. Nominally, that would be in developing bigger and more capable missiles for the military. But, as Korolev knew, and as he saw that von Braun also appreciated, a rocket that could send a Russian warhead high enough and far enough to drop it on America could also put an object into space and leave it there.

We have available only limited, and not wholly reliable, accounts of that critical time just after the war when the German rocket team's future was in the balance. There is a note von Braun reportedly prepared for his 'hosts' in which he wrote, 'The V-2 is an intermediate solution, which compares with future possibilities the same way a WW1 bomber compares to a modern bomber . . . A complete mastery of the art of rockets will change conditions in the world in much the same way as did the mastery of aeronautics' (M. Neufeld 2007). He went on to describe his vision for an orbiting, atom-bomb carrying space 'observatory' (M. Neufeld 2007).

We do know that von Braun worked under Korolev's direction for many years, years in which the Soviet Union had huge success in both the military and space exploration uses of rockets. Von Braun was never in the spotlight; this was not as much to hide his role from the Americans as from the Soviets' own population. Von Braun got into the news once: he was subpoenaed to testify at the Nuremburg War Crimes Trial of George Rickhey in regard to slave labour at Mittelwerk (Cadbury). More than one indignant US Congressman called for war crimes indictments against not only Wernher von Braun, but also his brother Magnus, who had worked full time at Mittelwerk. Neither brother was indicted, and the Soviets declined to make either available for testimony. Other than that, von Braun kept a low public profile. For that matter, neither was Korolev a public figure. Until after his death he was usually only referred to publicly as 'the Chief Designer', reportedly lest the CIA identify and target him for assassination (Scott and Leonov). If the CIA thought von Braun was the chief rocket man, that was all right with the Kremlin.

Shortly after the war, Korolev attended a Kremlin meeting where the Aviation Minister, Georgi Malenkov, discussed the new capabilities the Soviet Union needed. Malenkov noted that 'We are not going to fight a war with Poland . . . There are vast oceans between us and our potential enemy' (Heppenheimer). Korolev then briefed Stalin about the captured German rocket capabilities and their intercontinental missile design. Stalin was excited: 'Do you realise the tremendous strategic importance of machines of this sort? They could be an effective straitjacket for that noisy shopkeeper Truman. We must go ahead with it. The creation of transatlantic rockets is of extreme importance to us' (Heppenheimer). Stalin authorised Korolev to proceed and to use the German team as needed.

The Soviets 'invited' all of the German rocketry personnel under their control to take up residence in the USSR (Brzezinski). Trainloads of captured documents, V-2s and associated equipment headed east from the Soviet occupation zone of Germany. Some trains headed toward Podlipki, near Moscow, others toward Tyuratam in the Soviet Socialist Republic of Kazakhstan. In Podlipki, the Germans essentially re-established their rocket research and development facilities as a part of Scientific Research Institute NII-88, where Korolev was Chief Constructor. Near Tyuratam they set up launch facilities to test their improved rocket versions.

Some of Korolev's colleagues wanted the Germans to be fully dispersed and used in disparate Soviet facilities primarily to train young Russian technicians and engineers. That would be a good idea, he thought, if there had been only a few Germans with miscellaneous capabilities. But Korolev realised that he had available essentially the entire team that had developed rockets the likes of which the world had never before seen. True, it was important to have these Germans train Soviets (he'd have what would today be called 'interns' work alongside Germans), but it was also important to take advantage of the Germans' teamwork. He quickly saw that there was intellectual synergy among the members of the team that von Braun had spent years assembling and developing. The best way for Korolev to deliver the rapid results Comrade Stalin wanted was to keep the German team intact, working under his supervision.

He worked especially closely with von Braun, whom Korolev recognised as both a gifted designer and a kindred visionary. Their first objective was a missile with about twice the V-2's range, and improved accuracy. They would then develop larger missiles with greater range and warhead (or orbital) capacity. In part, this growth would come from increasing the size and power of the rocket's single engine. But, as von Braun's team had envisioned for his A9/A10 missile, the Russians agreed that the much larger vehicles would need to use multiple rocket engines.

While the rocket men envisioned rockets putting satellites into orbit, their masters in the Kremlin thought about being able to drop bombs on America. As noted earlier, in the immediate post-war years Stalin made every effort to conceal from the West the extent of the damage the war had done to his country; he did not want knowledge of his weakness to tempt his adversaries (Sheehan). The US had the bomb and the bombers to deliver it; the USSR had neither, but Stalin was determined to remedy that as quickly as possible. Resources were scarce, and were needed to rebuild the shattered economy. But Stalin assigned high priorities to developing both a Russian atomic bomb, and the means to deliver it. If the inexorable forces of Marxist-Leninist history were in his favour, Stalin would have an intercontinental bomber, or a missile, or both, ready by the time he had a bomb for it to deliver.

So Korolev got the industrial priorities he needed to build a super rocket. But in post-war Russia, even Communist Party mandates could not always summon scarce supplies into existence (Cadbury). Then there was the organisational friction. Sergei Korolev was not the only surviving Russian rocketeer, and colleagues such as Valentin Glushko, Mikhail Yangel and Vladimir Chelomei collaborated with but also competed against him. The bureaucratic jostling and infighting among the several rocket development 'bureaus' was not unlike the turf battles that raged routinely through the Pentagon. In 1946 Stalin had even set up a Special Committee to quell the wasteful squabbling and oversee a coherent rocket programme (Hardesty and Eisman). This committee did not eliminate competing organisations, but sought to channel their competition productively, without too much duplication and waste. It did not prevent intensely personal rivalries, such as between Korolev and Glushko, who had reportedly denounced each other during Stalin's purges (Cadbury). Some of the Germans were divvied up among the several bureaus, but the core von Braun team remained at NII-88 under Korolev.

The pace of the rocket development work was therefore set much more by the availability of tools, equipment and materials than by expertise.

Some official Soviet histories of these early days of the Space Race have recently become available, as have several unauthorised accounts, written by participants rather than by Party functionaries. In the official versions, the gifted and hardworking Soviet engineers and technicians developed a series of ever-larger rockets which invariably flew successfully from the first try, thus clearly demonstrating the superiority of Marxist-Leninist ideology. In these histories, if the Germans are mentioned at all, they are shown as having almost no substantive role. They appear like an admiring chorus in the background.

The unofficial accounts tell the story rather differently: over the years in Germany von Braun and his colleagues had developed an intense work culture focused on thorough quality assurance. They checked and rechecked every calculation, every specification. They closely inspected, at least twice, every fabricated part to ensure it met tight tolerances. That deeply ingrained quality-focused approach served them well in their new home. Between the urgency of Stalin's mandate and the programme's scarce resources, they could not afford to waste time or material on failed tests. The Germans preferred to perform all the critical tasks themselves. When Korolev insisted that Russian technicians do some of that work, von Braun showed him examples of shoddy workmanship on equipment and components fabricated by those Russians. There were sloppy assemblies, loose fittings, faulty wiring, even contamination in fuel lines. Von Braun explained what he and his colleagues had learned early in the V-2 programme: this new technology was far less tolerant of the workmanship imperfections with which a tank could still function or a plane could still fly. 'In the early days at Peenemünde before we learned this lesson, we blew up a lot of these buggers,' von Braun is reported to have told Korolev, 'and dropped even more into the sea.'

A sentiment attributed to von Braun goes 'It's not that we are geniuses; it's just that we old-timers have been working on these things so long, we've had years and years to make mistakes and learn from them' (Ward). Or, he may have laid it out more starkly: if Korolev insisted on letting lackadaisical Russian workers assemble the rockets, be they committed Party members or not, there would be potentially disastrous launch failures that Korolev would have to explain to Stalin. Did Korolev want to tell Comrade Stalin that an advanced rocket blew up because a sloppy worker let dirt get into the fuel line?

Of course, von Braun knew that his insistence on quality assurance could earn him and his team a new occupation – mining gold in Siberia (as Korolev had done for several years before the war). But then Korolev's rockets would blow up, and who would be next to board the train to Siberia? Von Braun also knew that his Soviet masters could turn him, and his brother among others, over to the Nuremburg War Crimes Tribunal for their role in war crimes at the Nordhausen/Mittelwerk complex (Cadbury). But von Braun was willing to risk his masters' displeasure in pursuit of his passion to put ever larger rockets into space, and he insisted that there be rigorous quality control in every step of the enterprise: 'If you don't follow our quality standards, you'll eventually get an intercontinental rocket – around 1960 or so, but you'll blow up a lot of rockets and maybe kill some good people before you do.'[2]

The poor-quality workmanship attributed to Russian workers in the missile programme in the early days (Cadbury) largely disappeared. It was still there, but von Braun had persuaded Korolev to institute rigorous quality-assurance procedures so that things like the backwards-wired switch got detected and fixed before it was installed in the missile.

Sergei Korolev was a gifted designer in his own right. He and von Braun developed professional respect for one another and worked productively together, as long as it was clear who was in charge. The collaboration was so close that it is impossible to say which features of which rocket were of German and which of Russian genesis. The R-2, the rocket that doubled the V-2's range, was truly a product of the hybrid team. It flew successfully 380 miles on its first launch, in October 1947.[3] In April 1949 another hybrid missile with German and Russian design lineage flew over 1,200 miles. It was becoming apparent to the scientists, and to the Soviet leadership that an intercontinental-range missile was within their grasp in the next few years. As soon as it was available, the intercontinental ballistic missile (ICBM) would remedy the dilemma the Soviets then had: they now possessed the bomb, but no way to deliver it beyond Europe.

The atomic bomb the USSR had successfully tested the previous August (1948) was an implosion-type device closely resembling the US Fat Man device. That was no coincidence, because Soviet spies such as David Greenglass, Klaus Fuchs, Ted Hall and Oscar Seborer had stolen much of the design from Los Alamos (Sheehan; Broad). Like Fat Man, the Soviet device weighed about 5 tons, and could only be carried by Tu-4 bombers, the reverse-engineered copies of the Americans' B-29s that had been forced to land in Siberia during the war (USAF 2017).

---

2. See 'What Really Happened' (23 Oct 1960).
3. See 'What Really Happened' (Sep 1949).

Like the 'Silverplate' B-29s, the Soviets' version of an atomic-capable Tu-4 bomber was reconfigured to be able to haul their nuclear weapon. However, these modified bombers would be of limited strategic importance: they did not have the range for a round trip to the US.

In the short term, the Soviets would try to maintain uncertainty about their capabilities, such as by making references to aerial refuelling capabilities for their bombers. They hoped to shake the Americans' (accurate) belief that the Soviets had not yet mastered that technology. The Soviet premise was that while actual military capabilities were important, what your adversaries believed those capabilities to be could be equally important. Yet the Kremlin knew that the Americans were no strangers to disinformation. (The US Army, for example, had convinced the Nazis that George Patton would launch the main invasion of France at Calais with the 'First US Army Group', which consisted entirely of paper, faux radio traffic and lots of inflatable tanks and trucks parked in motor pools around Dover, facing the Pas de Calais [Holt].)

So the Soviet leadership realised they likely could not deceive the Americans for long about their delivery capabilities. It was imperative that the Soviet Union truly acquire a nuclear delivery system as soon as possible. To that end, Sergei Korolev received the mandate to proceed at the highest priority to develop a missile capable of delivering a 5-ton warhead to a range of 5,000 miles (Hardesty and Eisman). Thus, in October 1949 Korolev and his 'junior partner' von Braun had a green light to pursue their mutual dream of building a rocket capable of launching an earth satellite. To be sure, it would meet the military's specifications for an ICBM with an 11,000lb warhead. If it were carrying a 3,000lb payload instead, and was therefore able to reach a greater velocity, this missile would be fully capable of putting that payload into orbit (Heppenheimer). Their new rocket would be like the optical illusion that appears to some viewers to be a well-dressed young woman, while to others the same image shows a witch in a shawl.

At nearly twice the height and power of the V-2, the 1,200-mile R-5 missile that Korolev had tested in late 1949 was still a single-stage derivative of the German rocket (Wade).[4] To get intercontinental capabilities would require more than just making every component bigger. In part that's fundamental physics: a horse 'scaled up' to be as tall as an elephant would collapse with broken legs. An elephant's leg bones are not just longer, but proportionately more massive than those

---

4. See 'What Really Happened' (15 Mar 1953).

of a horse, as they need to be to sustain the much greater mass of the elephant's body. Analogously, the new 'R-7' ICBM would have to be not just bigger, but structurally much stronger than its predecessors. This leads to the engineering conundrum: additional structural weight requires the rocket to generate more lift, which requires additional fuel, which adds yet more weight, requiring more structural strength, and so on.

The rocket pioneers Tsiolkovsky in the Soviet Union, Goddard in the US and Hermann Oberth in Germany had all invoked the use of multiple rocket stages to help deal with this problem of structural weight (Curtis). A rocket uses a lot of fuel just to get off the ground. Once that initial portion of fuel is gone, the tank that contained it is useless weight. Rather than expend fuel on continuing to carry that useless weight higher and faster, 'multi-stage' rocket designs call for the now-empty section to be dropped off. Also typically dropped are the heavy rocket engines that were initially needed to get the whole thing off the ground. A second stage, with its own fuel supply and smaller engines, then kicks in and further lifts the now-lighter vehicle. There can be additional, successively smaller stages. As noted earlier, von Braun's wartime designs included a two-stage intercontinental missile, and three- or four-stage orbital launch vehicles.

Developing more powerful rocket engines for larger rocket vehicles was also a challenge. This was more than just a matter of enlarging and proportionately bulking the engine up (the elephant vs. the horse). It is essential that the combustion of rocket fuel occurs evenly: irregularities in the burning create uneven pressures, which can damage the engine. The larger the engine's combustion chamber, the harder it is to ensure that all the fuel in it is burned evenly throughout. So at some point, instead of increasing the size of a single engine, it makes more sense to use multiple engines side by side to get increased thrust. But that not only adds weight, but also complexity, which increases the chance of a failed component.

Still, all this was exactly what Wernher von Braun, Sergei Korolev and dozens of their colleagues lived for – to work through and solve the challenges. How many engines could be put together? How big could they be? How to make sure they would all work when and as needed? How to size the stages and how many with how many engines? And of course, how to guide the payload accurately to its target (be it New York or earth orbit)? The work proceeded well, if not always quite smoothly. There were disputes among the great Russian egos such as Korolev, Glushko, Yangel and Chelomei. Fuelled in part by personal animosities, these disputes involved critical matters like engine design and choice of fuel. Sometimes too they were merely the kind of bureaucratic turf battles that are the modern manifestation of tribal warfare (Cadbury).

There was also tension about the Germans. None of the Russians wanted to admit how valuable the German scientists, engineers and technicians were. Von Braun and most of his top men still worked under Korolev at the NII-88 bureau, although a few Germans had been assigned to work at Glushko's and other rocket design bureaus. (Although all state-owned and controlled, the several design bureaus operated largely but not totally in competition, not unlike Boeing vs. Lockheed in the US.) By the early 1950s, many of the Russian rocket men believed they had gotten what they could out of the Germans and wanted Soviet rocketry to be unmistakably a Russian enterprise. Korolev shared this view in large part. He sent many Germans back home, or at least as far as East Germany (T. Reed). But Wernher von Braun and dozens of others would have had to be forcibly taken back to Germany. In the Soviet Union they were pursuing their passion for space. They lived reasonably well (and probably far better than they would have back in the Fatherland), and they believed that the rockets they were building would help write the history of space exploration (if they didn't help wipe out civilisation first). Korolev shared that cosmic aspiration and, as long as the Germans stayed in the background, he was content to let them continue to help him.

The R-7 development progressed in the early 1950s, although perhaps not as quickly as it might have if the Russians weren't squabbling among themselves. It might have gone faster and more smoothly too if the original cadre of Germans had not been thinned out. Nevertheless, as of March 1953, the R-7 was slated to begin flight tests at the Tyuratam test range early the next year.[5] In March 1953 Korolev believed that Stalin was satisfied with the ICBM programme's progress. He was stunned to learn of Stalin's death.[6]

---

5. See 'What Really Happened' (Mar 1954).
6. See 'What Really Happened' (5 Mar 1953).

Chapter 22

The Rocket Men – Americans

At the end of the war, Caltech's Dr Theodore von Karman headed a team of aeronautics experts. They were tasked to advise the Army Air Force on the future of US air power (Mindling). Von Karman's group travelled to Europe, interviewed German scientists and engineers in the western sectors of occupied Germany, and collected German equipment and documentation from aircraft and engine factories and from aeronautics labs (Mindling; Sheehan). Von Karman later said he was astounded by the progress German aviation had made and was 'thankful that Hitler and his generals had failed to take full advantage' of this work (Sheehan). The von Karman team had access to the limited V-1 and V-2 material the US Army had retrieved from the Mittelwerk complex. To their regret, what they did not have access to was the team of rocket scientists who had developed the V-2 rocket weapon.

Von Karman was concerned that most of the German rocket developers were unavailable to the US. 'I'd feel a lot better,' he told a colleague, 'if we had kept the German rocket scientists out of the Russians' hands. Maybe we wouldn't have let them do very much, perhaps just put them out at some remote base out in the desert and let them putter around, as long as our guys could pick their brains once in a while.'[1]

Since he didn't have the Peenemünde team available, von Karman decided to solicit advice on the future of US missiles from Robert Goddard, America's own top rocket scientist. Goddard had been doing pioneering rocket research since the 1920s; he held numerous patents on rocket components. His publications were widely read and respected abroad, including by the Germans. The story is told that when a captured Luftwaffe general was being questioned about some of the German technology, he replied, 'Why don't you ask your Dr Goddard?'

---

1. See 'What Really Happened' (mid-1950).

(Heppenheimer). Yet Goddard and his work were largely ignored by the American aeronautics community (J. Neufeld). He had not initially been on von Karman's study team. There was some professional bad air between the two from before the war: Goddard had a penchant for not sharing his findings with other scientists, as is the general custom. This had annoyed von Karman (Burrows). Or perhaps the problem was that during the war Goddard was doing rocket research for the Navy, not the Army as von Karman was doing (Lehman). At any rate, Goddard was in ill health at the end of the war and could not have made the European trip even if he'd been invited. By then, however, he had been looking into German rocketry on his own.

From pre-war correspondence with various German rocket scientists Goddard knew that the German government was far more interested in rocketry than was the US (Emme). Then, right after the war, his Navy work at Annapolis, Maryland, gave him the opportunity to inspect the German rocket material that US troops had taken from Mittelwerk (Ordway and Mitchell). His initial reaction to reviewing the German material was that 'it looked a lot like his rockets' (Heppenheimer). He concluded that the von Braun team had stolen a lot of his own work (Lehman). From what he could see of the various V-2 pieces and documents, Goddard then realised that the V-2 was technically far more advanced than the most successful of the rockets he had designed and tested (Lehman).

His further analysis was that the Germans who developed the V-2 were clearly capable of creating bigger rockets, with more lift and more range, and that doing so might not take them very long. Goddard said that he would not be surprised if von Braun and colleagues already had designs for intercontinental-range rockets. (As noted above, unknown to Goddard, von Braun had indeed begun development of a missile to hit New York. Moreover, it has emerged that Goddard's early work was inadvertently helpful to the Germans. In post-war writings now available in the West, von Braun stated: 'Goddard's experiments in liquid fuel saved us years of work, and enabled us to perfect the V-2 years before it would have been possible' [Levine].)

Goddard believed that if, as the intelligence community assumed, the German rocket team was intact and working for the Soviets, the combination of Russian and German rocket expertise could present the US with a serious threat in just a few years. Goddard himself had benefited greatly in his work from the extensive theoretical work of the Russian Konstantin Tsiolkovsky, who had pioneered many aspects of astronautic theory in the early decades of the twentieth century (Anonymous 2004).

Goddard assumed therefore that the Russian rocket community itself might be as technically advanced as the von Braun team. The fact that the Soviet Union had not deployed a weapon like the V-2 of their own during the war was no more surprising than that the US had not done so – it was a matter of allocating priorities and resources. Hitler had poured enormous effort into trying to terrorise his enemies with Vengeance weapons; his enemies concentrated on destroying the Reich's military capabilities. 'I wouldn't be at all surprised,' reflected Goddard, 'if the Soviet equivalent of the Peenemünde gang had been pretty close to their own V-2 when the war ended.'

It wasn't. In 1945 Goddard had no way of knowing that much of the Russian rocket science community had been destroyed and imprisoned in 1938 during Stalin's purges (Ward). Given the information available to the West, Goddard's concern about the capabilities of Russian rocketeers and the dangerous synergy with their German counterparts was well founded. Goddard would presumably have been even more concerned if he had known that some of his own pre-war rocket research for the US Navy had been stolen by the KGB (Haynes et al).

He had not known the German rocket men before the war on a personal level, but Goddard reasoned that as scientists and engineers, they would be more concerned with advancing astronautics than with their masters' political systems. Besides, he reflected, these Germans had already spent years developing deadly weapons in support of the murderous Nazis. So Goddard believed the Germans were likely to work more or less willingly for the Soviets.

Goddard was dying of cancer, but he put together a report for von Karman's study team to submit to the Army Air Forces. In it he advocated, not for the first time in his life, that the US step up its rocket research (Goddard and Pendray). He urged the Pentagon to vigorously pursue developing the kind of powerful, militarily-significant rockets which he believed the Soviets, with German help, could have in a just a few years. In a technical appendix to his report, Goddard provided a list and brief review of his forty-eight pre-war patents and of the thirty-five more patents issued to him during the war, noting that the Germans had evidently found many of them useful. He suggested with too-obvious sarcasm that the US military might now find them useful. This appendix achieved wide circulation within the aerospace community for many years thereafter, helping post-war rocketeers to make full use of Goddard's previously unappreciated work.

In the months right after the end of the war, von Karman's team submitted a report to General Henry 'Hap' Arnold, chief of the Army

Air Forces (Mindling). The thrust of the report was that the future strength of the Air Forces (and therefore the future security of the US), would depend on continuous scientific and technological progress, and swift adaptation to new developments. It would be a constant race to invent what could be invented before the other guy did so (Sheehan). In addition to discussions of future aircraft such as supersonic jets that could deliver atomic weapons from great distances, the study also envisioned unmanned bomb-carrying aircraft more capable and accurate than the Germans' V-1. There were also discussions of rocketry (Wattendorf et al.). The limited range and accuracy of the V-2 were noted, but with recognition that far more powerful and more capable rockets based on these demonstrated technologies were foreseeable. Von Karman wrote in an introductory volume: 'Rocket research and development have become one of the most important responsibilities of the Air Forces for the future' (von Karman).

A key team member, Hugh Dryden wrote:

> The experiences in the tactical use of guided missiles in this war indicate that another war will probably be opened by the descent in large numbers of missiles launched from distances of perhaps 1000 to 3000 miles . . . It is vital to the future defence of our nation that research and development in this field be continued not only so that adequate countermeasures may be developed against enemy missiles but also that we may have available the best weapons of this type in the world (Dryden et al)

Insofar as von Karman's Scientific Advisory Group was calling for better bombers and associated technologies, they were preaching to the Air Forces' choir. Most of the senior leadership of the Army Air Forces (e.g. Hap Arnold, Tooey Spaatz, Curtis LeMay among others) were bomber pilots and advocates of air power, which, many claimed, had made critical contributions to the defeat of Germany. Outspoken generals such as Curtis LeMay insisted that the Air Force needed to concentrate on manned bombers as the key strategic weapon of the future. They (selectively) cited data from the massive survey report on how much German military capability had been affected by the USAAF and RAF bombing campaigns (Strategic Bombing Survey). These Air Force generals believed it was imperative that the advances in aircraft technology which had begun to emerge during the war be developed right away. The existing B-29 was a potent weapon, but it did not have the truly intercontinental range that could be needed next time.

> ## WHO REALLY WON THE RACE?
>
> According to an account that has now become available, the Germans thought they were in a rocketry race with the US. Despite Goddard's secretive habits, the Germans had studied his pre-war work, concluding from it that large, long-distance rockets were feasible. An unclassified 1939 article published by a US Army major advocated a '3000 mile missile' (Mitchell). Therefore the Germans logically assumed that the Americans would be pursuing this technology themselves.
>
> When they learned during the war that Goddard had seemingly disappeared from his New Mexico research facility, they took this as indicating that Goddard was heading up the secret American rocket programme. Using this intelligence information, von Braun's boss General Walter Dornberger convinced Hitler that they were in a rocketry race against the US. Dornberger thereby secured top priority of resource allocation for the missile work at Peenemünde (M. Neufeld 1996).
>
> Goddard did leave New Mexico to work for the military, but despite his recommendation to Washington that they should pursue large rockets, the work the Navy hired him for dealt only with small rockets to help heavily-laden aircraft take off (Lehman). So the US was not in the race the Germans thought they were running. But because of their misperception, the Germans diverted enormous resources to their missile programmes. Though the V-1s and V-2s killed thousands of civilians, they did not generate the terror that the Germans had hoped would make a strategic difference in the war. By one estimate (Ordway and Mitchell), the German missile project cost them 50 per cent more than the Manhattan Project cost the US. For the cost of one V-2, the Nazis could have built six fighter planes to battle Allied bombers (Stoker). Or they could have made more of an effort to produce the Wasserfall anti-aircraft missile, a smaller version of the V-2. According to Albert Speer, the Reich's Armaments Minister, deployment of the Wasserfall would have stalled the Allied bombing offensive (Speer). It is left for 'Alternative History' speculators to game out how that would have changed things.

Indeed, several such 'Research and Development' efforts were already underway while the war still raged. In light of von Karman's report, General Arnold and others were all the more confident that these wartime R&D efforts needed to continue. Because of these attitudes in

1945, the decidedly odd B-36 first flew in 1946, with its six piston and four jet engines; the B-50, an upgraded B-29, flew in 1947, as did the first all-jet bomber, the B-47 (Knaack). Meanwhile, design for a more-capable jet bomber, the B-52, had started in 1945. The consensus was that these several bombers' capabilities to deliver atomic weapons to far-distant targets would be an important strategic element in preventing the next war, or winning it.

General Arnold sought to make sure the Air Forces' weapons would always represent the most advanced capabilities then possible. He wanted there always to be 'prodigies of futuristic thought' on the drawing boards (Sheehan). Accordingly, in the months right after the end of the war, he set up a Research and Development office and assigned Colonel Bernard Schriever to lead it (Sheehan). Schriever, a former test pilot with a master's degree in aeronautical engineering from Stanford, was wholly receptive to Arnold's counsel on the importance of scientists for the future of air power. The next war would be decided by brainpower, not brawn, said Arnold (Sheehan).

Arnold and Schriever promptly set up numerous new research projects with monies that had been appropriated but not yet spent when the war ended so suddenly. They knew that Congress would soon cut back on military spending, so there was a brief opportunity to get as much money as possible committed to contracts for developing the next generation of weapons. In 1945 and early 1946, there were funds sufficient for efforts besides the new bombers mentioned above.

Among these were several missile projects, including research into unmanned jet aircraft capable of flying long distances carrying atomic bomb payloads. Such air-breathers would be called 'cruise missiles' today. There was research into missiles to intercept other missiles (Baucom). There were also studies on artificial satellites. Right after the war, the Navy approached the Army Air Force about collaborating on research into developing an artificial satellite for weather and photo reconnaissance (M. Neufeld 2007). The Air Force declined to work with the Navy, although they did contract Douglas Aircraft to conduct a satellite feasibility study (M. Neufeld 2007). This bore fruit several years later.

Also included in the Air Force's R&D programme right after the war was a contract to study the feasibility of an intercontinental ballistic missile capable of carrying a large warhead (Parsch 2005). The Consolidated-Vultee Aircraft Company (Convair) won that small study contract and put Belgian-born Karel Bossart in charge of the team. Bossart proceeded to design a rocket, dubbed 'Hiroc', which was intentionally very different from what he knew about the Germans' V-2. He had in mind advances

> ## Going Ballistic
> The jet-powered 'missiles' mentioned above have wings and fly like an aircraft through the atmosphere on a roughly level course to their target. By contrast 'ballistic' missiles are more like artillery shells, fired upward in a high, curved trajectory so that they will fall onto their target. While artillery shells reach heights of several miles, rockets such as the V-2 reached heights of more than 50 miles before hitting the ground less than 200 miles from its launch site. Intercontinental-range missiles would spend most of their flight hundreds of miles above the earth, i.e. in space.

in regard to the fuel, the fuel tanks, and the engines (Sheehan). For example, his design called for aluminium fuel tanks rather than steel, thus saving a great deal of weight (Heppenheimer). The record does not indicate whether this was purely an effort to advance the state of the art, or else may have had some element of anti-German sentiment. It could even be the classic bureaucratic 'Not invented here' hostility to other organisations' ideas.

While these several studies were underway, Colonel Schriever, in the Air Force's R&D office, digested von Karman's report. Although its major emphasis was on bombers, Goddard's section in particular exhorted prompt attention to the threat posed by the combined rocket expertise of the Russians and the Germans. Drawing on US and UK sources, Schriever delved into the available information on the V-2 and its effectiveness. Its range was only 200 miles, it was very inaccurate, and it carried a payload of about 2,000lbs of explosive. Its strategic impact on the course of the war had been negligible. Moreover, since each rocket was a single-use item, it was an expensive, imprecise way to deliver a one-ton bomb (J. Neufeld). In that regard, it compared poorly to an aircraft that could potentially fly many missions, dropping tons of bombs each time at a far longer range. Churchill's wartime science advisor Frederick Lindemann, later Lord Cherwell, had found it hard to believe that the Germans had 'squandered resources on such an ineffective weapon' (Heppenheimer). As noted above, perhaps Robert Goddard unknowingly fostered the Germans' 'misallocation' of resources.

However, Schriever also realised the inherent advantages of a missile such as the V-2. General Pile, head of UK anti-aircraft defence, wrote that 'by far, the most difficult problem we ever had to tackle was to find some means of defence against the V-2 rocket' (Stoker). Falling to its target from

50 miles high, and travelling at more than 3,500 miles an hour, the V-2 was an essentially unstoppable weapon. Capable of attacking not only in daylight but under all weather conditions, it was far too fast for fighter aircraft or even radar-guided anti-aircraft guns to intercept (McMurran). It arrived without warning – travelling faster than the sound it made. It was expensive in material and production costs, but its use did not involve any potential loss of highly-trained aircraft flight crews (Stoker).

Out of about 3,200 V-2s fired at Great Britain, there is only an unconfirmed, and highly unlikely, report of a single one ever being destroyed in flight – by machine-gun fire from a B-24 bomber returning from a raid on the continent (Rosenberg). By contrast, manned bombers were shot out of the sky by the thousands during the war.

True, a long-range missile lacked the flexibility of a piloted aircraft in regard to reaching or even changing its target. Yet it would have far greater assurance of being able to reach its destination. Since well before the war, advocates of airpower (sometimes called the 'Bomber Mafia' in the US), had echoed the thought ascribed to Britain's Stanley Baldwin: 'the bomber will always get through' (Baldwin). The experience of the war had shown that this was only true in the plural. With enough bomber aircraft, some will always get through, albeit at terrible material and human cost. But now in regard to V-2-like missiles, the prospect was that 'every missile will get through'.

Schriever read that even during the 1944 V-2 campaign, Dr R.V. Jones, science advisor to the UK Air Ministry, had written:

> There can be no doubt that the [V-2] rocket has come to stay . . . With a very long range rocket . . . it may be easier to increase the radius of destruction by the use of new types of explosive based upon the fission of the uranium atomic nucleus . . . A two stage rocket could deliver a 1 ton warhead to nearly 3000 miles, with a probable error of 10 miles in range . . . This might be a feasible weapon for delivering a uranium bomb, should such a bomb become practicable (Stoker).

Also, in November 1944, Duncan Sandys, in charge of the task force on V-2 countermeasures, wrote:

> The flying bomb and the rocket used against us in 1944 are of course only the fore-runners of other long distance bombardment weapons of this kind. Given time, there can be little doubt that the effectiveness of the existing [V-2] could be appreciably

improved upon.... In future the possession of superiority in long distance rocket artillery may well count for nearly as much as superiority in naval or air power (Stoker).

Schriever found that some American military experts were of similar mind. In May 1946 General Joseph Stilwell wrote in a report for George Marshall on the equipment the Army of the future would need:

> Guided Missiles, winged or non-winged, travelling at extreme altitudes and at velocities in excess of supersonic speed, are inevitable. Intercontinental ranges of over 3000 miles and payload sufficient to carry atomic explosives are to be expected ... Such missiles would be incapable of interception with any existing equipment such as fighter aircraft or antiaircraft fire. Guided interceptor missiles ... will be required (War Department Equipment Board).

As is often the case with innovations in science and technology, opinions within the community of technical experts were mixed. Dr Vannevar Bush had headed the National Advisory Committee for Aeronautics (NACA) before the war, and FDR's Office of Scientific Research and Development during it. Before a Congressional committee a few months after the war he opined that a ballistic missile that could carry the 5-ton weight of an atomic bomb was 'at least ten years away' and that he didn't think anyone in the world knew how to deliver an atom bomb by ballistic missile. 'I feel confident' he said, 'that it will not be done for a very long time to come' (Mindling).

Dr Bush's credentials notwithstanding, Schriever reasoned that with the combined expertise of the Soviet and the German rocketeers, the USSR might substantially shorten his 10-year timeframe for producing a nuclear-capable intercontinental missile, especially if the V-2 rocket team had continued their work essentially without pause in their new Soviet digs. 'The Soviets may be way ahead of us with rockets, and getting further ahead every day' thought Schriever.

He further reckoned, as Stalin himself had done, that the Communists' geographic asymmetry with regard to the US was a potential incentive for the Soviets to put a very high priority on long-range missile work. Like his Soviet opposite numbers, Schriever could see that unlike the US, the USSR had no place from which any bombers they might have could threaten the America with an atomic attack.

For that matter, how quickly would the Russians get an atomic bomb? Without yet knowing the full extent of Soviet penetration of the Manhattan Project, General Groves had projected it would not be until sometime in the 1960s (Sheehan). Other military sources thought it would not be until 1953 at the earliest, while the British estimated 1954 (Aldrich). In any event, it would possibly be less than 10 years. Thus, Schriever reasoned, the Soviets could be making an all-out effort using their newly-augmented rocketry expertise to ensure they had a missile capable of delivering atomic bombs just as soon as they had some of them. As discussed above, that was exactly Stalin's reasoning in pushing Korolev and his Russian-German team.

By January 1946 Schriever convinced General Arnold that aggressive research into long-range rocket feasibility would be prudent. Even if the Air Force decided to continue its reliance on improved manned or unmanned bombers, it was nevertheless important for the US to determine what the other side might be able to do with long-range rockets, and how quickly. Arnold agreed that he didn't want the US to have only a fleet of nuclear-capable bombers that could reach the Soviet Union in a matter of hours, while the 'Russkies' had a fleet of nuclear-tipped missiles that could reach the US in a matter of minutes. That would be, he thought, like bringing a knife to a gunfight. Nor did the Air Force assume that they could soon develop anti-missile interceptor missiles that would reliably neutralise any incoming missile threat.[2]

Arnold retired from active duty in late February 1946 (Fogerty). Just before then, he directed Schriever to amend the Hiroc missile research contract with Convair, to shorten the schedule, increase its payload, and its budget.[3] Originally this project called for the launch in 1955–6 of a missile carrying a 5,000lb warhead to a range of 5,000 miles (Neufeld 1990). Schriever's revised specifications increased the warhead to 8,000lbs, with the first operational missiles to be not just tested but operationally deployed by the end of 1955.

The contract's original 10-plus years of development would have been in keeping with Vannevar Bush's prediction. But as Schriever expressed it with intentional wryness, 'This program needs to be fast, not half-fast'. His model was Leslie Groves and the Manhattan Project (J. Neufeld). 'If we can invent and build an atom bomb and all the associated reactors and processing plants in three years from a standing

---

2. See 'What Really Happened' (1946–2020).
3. See 'What Really Happened' (Jan 1946).

start, surely with the head start of having a lot of the V-2 design in our hands, we can build a much more capable rocket in that sort of timeframe. It is a good bet that Ivan – and Hans – are working together on doing just that right now.'

The Convair team responded positively to the higher priority, and the bigger budget. They were given access to all of the captured V-2 parts and documents (Lethbridge) and were able to hire a few of the German rocket technicians who had been in western Germany when the war ended. Design proceeded quickly through the rest of 1946 and into 1947, with planned launches of scaled-down test vehicles in 1948.

By 1947, Pentagon leaders could hear the sound of budget axes being sharpened on Capitol Hill. General Arnold's successor Tooey Spaatz was a long-time bomber supporter. He was loath to give up any of the several piloted bomber projects (the B-36, B-47, B-50 and the B-52). Besides the Hiroc ballistic missile there were several other unmanned systems in development. These were V-1-like cruise missiles of various ranges and speeds (subsonic vs. supersonic) such as the Navaho, the Matador, the Snark and the Boojum (J. Neufeld). Such missiles were attractive because they could have the range and the heavy-lift capacity needed for nuclear payloads (Mindling). Unlike re-usable manned bombers that needed to return to safety, cruise missiles' entire range could be devoted to a one-way trip, nor did they need room aboard for a crew. On the other hand, while their jet engines would propel them faster than wartime piston-driven bombers, it would still take hours for these air-breathers to reach targets in the Soviet Union, even at supersonic speeds. Moreover, like the V-1 before them, they would fly slow and low enough to be vulnerable to enemy anti-aircraft defences.

By contrast the Hiroc ballistic missile would travel almost ten times as fast and would presumably be as unstoppable as the V-2 had been. It was not at all certain how soon it could be built with enough lifting power to deliver an atomic warhead at intercontinental range, but the assumption was that the Soviet/German team was racing toward that same goal. There was little doubt that 'those other guys' would be successful sooner or later in developing such a weapon system.

Paraphrasing Hap Arnold, Colonel Schriever said 'To prevent the next gunfight, we need to have a gun'. He persuaded General Spaatz to maintain robust funding for the Hiroc, at the expense of funding for the several cruise missile projects.[4]

---

4. See 'What Really Happened' (Jul 1947 and Sep 1951(1)).

In 1947 and 1948, technical work on the Hiroc, now named the Atlas, was proceeding well. But organisational friction slowed the project. There was bureaucratic sniping from the traditional bomber advocates within the US Air Force (USAF), by then a separate service from the Army. There were also internecine turf battles among the several Air Force organisations and review committees who had their hands in R&D (Sheehan). Added to that were disputes with the Army about 'roles and missions'. The Army wanted to retain sole jurisdiction over all ground-launched missiles (McMurran).

After several successful tests of Atlas components in early 1948 Schriever proposed a streamlined organisation to ensure timely fielding of the system. He described the tangled web of dozens of reporting and review requirements of the USAF and the Defense Department. They were hampering the timely development of the ICBM (Sheehan). Careful not to denigrate the importance of manned bombers, Schriever noted that it was also important that the US keep pace with, and ideally win the race against, the German-augmented Soviet ICBM team.

Predictably, the Air Force Chief of Staff planned to set up a committee to study whether there were too many committees overseeing the missile programme. But before he could do so, the news broke in August 1948 that the Soviets had detonated a nuclear device – five years sooner than anyone in the US expected.[5]

As described previously, President Truman calmed public concern (and helped his re-election that autumn) by promptly reassuring the nation that America's atomic arsenal was far more advanced than the Soviets'. He further noted that 'Recent tests of a long range missile design have now demonstrated that we will have, within a few years, powerful rockets of global reach with which to defend our nation's security.'

In keeping with Truman's promise, Air Force Chief Of Staff Hoyt Vandenberg tore up the charter for his 'Committee to Review the Committees' and promptly signed off on the streamlined organisation and reporting structure that Colonel (soon Brigadier General) Schriever had proposed. 'Get it done' was the explicit mandate, 'before the Russkies do' was the unspoken part; '. . . at the lowest possible cost' was not part of this urgent mission directive.

Schriever set up the Western Development Division to run the ICBM programme from the Los Angeles area, where he could escape the bureaucratic friction in the (overcrowded) Pentagon and could more

---

5. See 'What Really Happened' (29 Aug 1949).

readily kick the nearby contractors' behinds when needed. Also to maintain control and expedite progress, he set up parallel development efforts for major missile systems using separate contractors. That way, if one contractor's components progressed inadequately, Schriever could have the Atlas use the materials from the second contractor. This was highly effective in stimulating every contractor to constantly produce their best work. It was also hugely expensive (Sheehan).

Despite the rapid progress, the cost of the Atlas ICBM caused internal opposition. General Curtis LeMay, head of the Strategic Air Command, took every opportunity to criticise the immense expenditures being devoted to a new technology that he considered would never be as reliable as his growing fleet of piloted bombers. LeMay was adamant that his B-29s, B-36s and B-50s would be fully capable of destroying any nuclear-capable Soviet bombers on the ground (Sheehan). Of course, this assumed that the Soviet fleet would still be on the ground by the time the US bombers arrived. LeMay didn't consider Soviet ICBMs to be a credible threat. The Pentagon and Congress listened to him, but not that much. He was happy to tell all who would listen how effective his incendiary bombing campaign against Japan would have been, if only

LESLIE GROVES' UNFINISHED PROJECT

When Leslie Groves was tapped to run the Manhattan Project – to build the Bomb – he was only a few months into another major project – the construction of the new War Department building on the Potomac River floodplain in Virginia. That unusually-shaped building, designed to fit an unusually-shaped site, came to be called the Pentagon, with its five sides, and five rings, and four above-ground floors. While it was still under construction in 1942, but after Groves had been diverted to the Manhattan Project, the War Department determined the building wouldn't be big enough. There was talk of adding a fifth floor to provide the additional space; the reinforced concrete structure could well sustain a fifth floor (Vogel). But without Groves, and his deputy Robert Furman whom he'd taken with him to the Manhattan Project (Roulo), the Pentagon project had fallen behind schedule and was far over budget. The delay and the cost of ripping off newly-installed roof sections to add an additional floor were soon determined to be too great. So the Pentagon was completed in its four-storey configuration, much to the chagrin of the many thousands of somewhat cramped personnel. (See 'What Really Happened' [Jul 1942].)

the atom bombs hadn't ended the war before it really got going. LeMay got some funding for his new B-47 and B-52 bombers, but not nearly as much as he wanted. He coveted Schriever's budget, but was never successful in raiding it. And the Atlas proceeded.

After the mid-1948 successful launches of test vehicles, the design of the full-sized Atlas proceeded well during 1949–50. The design specifications called for putting an 8,000lb warhead within 1,500ft of a target 5,000 miles away, equivalent to sinking a hole in one, right into the cup, from a mile away. Convair's proposed rocket would be 160ft long and 12ft in diameter (Sheehan). It would weigh 670,000lbs, (equivalent to five loaded B-29s, then the world's largest aircraft) and would be powered by seven engines delivering a total of 840,000lbs of thrust (Heppenheimer). It would be by far the most massive object ever to leave the surface of the earth, except perhaps for whatever the Soviets were putting together. On that point there were intelligence reports suggesting that the Soviets were testing a rocket engine about twice as powerful as anything the US had on the drawing boards (J. Neufeld). At that time the Soviets did indeed have a design for their R-3, a 3,000-mile rocket using one or more 260,000lb thrust engines to deliver a 3-ton payload (Heppenheimer; Zak (2)).

Fabrication and testing of components was underway in late 1951, a planned launch of the full-size vehicle from Florida's Cape Canaveral would occur in late 1953 or early 1954, and operational deployment of the first squadron of missiles (a few dozen) was slated for mid- to late-1955. Vannevar Bush's prediction that 'long range missiles would not be possible for a very long time to come' had turned out to be less than 10 years. In his later writings, General Schriever reflected on how things could have been different: 'Von Braun and the other German rocket men being in Soviet hands scared us spitless. It made us work a lot faster, I believe, than we would have if we'd had the Germans over here.'

In November 1951 the US detonated the world's first hydrogen bomb.[6] This first fusion device was not a deliverable weapon; it weighed over 80 tons (Herken 2002). That was twenty times as heavy as the warhead then planned for the Atlas. But a few months after that first demonstration Schriever learned from key Manhattan Project scientist John von Neumann that H-bombs weighing in at about 2,000lbs or less were likely to be available in about six or seven years (Sheehan).

---

6. See 'What Really Happened' (1 Nov 1952).

The huge Atlas ICBM that would be operational in 1955 had been designed to carry 8,000lb atomic bombs. It now appeared that much smaller nuclear devices would be available by about 1958–9. A concept therefore arose in early 1952 that perhaps the Air Force should redesign the Atlas as a smaller missile, and defer its deployment until smaller warheads became available. As noted, however, the CIA reported that the Soviet programme was making good progress on what might be missiles far larger even than the Atlas as now planned. And Soviet ICBMs were perhaps only two or three years away, the intelligence community said. Thus, it might already be unavoidable that the Russians would field some ICBMs before 1955, before the US could do so. That would be bad enough; but the Pentagon knew they needed to keep any such 'missile gap' as brief and as small as possible. They could not risk prolonging the missile gap by downsizing the Atlas and delaying it until smaller warheads became available.[7]

Added to this was the consideration that the Soviets' anti-aircraft defences against America's manned bombers were improving, according to the CIA (J. Neufeld). Russian surface-to-air missiles (SAMs) were getting more capable and more plentiful. The potential diminishing utility of bombers (i.e. fewer of the bombers would get through) was part of the argument that Dr von Neumann made at this time for a maximum US effort on ICBMs rather than bombers (Sheehan). He also argued that it would be logical for the Soviets to have a similar focus. The Russians had never had a long-range bomber capability, but they'd had substantial pre-war rocketry expertise, which was now enhanced with German input. So it would make most sense for them to rely heavily on a missile force for their post-war military posture rather than on bombers. The Communists were doing an end run around the Air Force's bombers, von Neumann said (Sheehan). The US edge in bombers was a wasting asset. The race with the Soviets was in missiles, not aircraft.

Therefore, in mid-1952, the Pentagon directed Schriever to begin parallel development of a second ICBM to be ready for deployment shortly after the original Atlas. This one would be far less massive (one-half or even one-third of the size of the original, the 'Atlas A') because it would only need to heft a one or two-ton instead of a four-ton warhead. A lighter payload would also allow it to have a have greater range, putting more of the Soviet Union within striking distance. This new ICBM would be a scaled-down derivative of the Atlas as

---

7. See 'What Really Happened' (Early 1953).

originally designed. This demi-Atlas, or Atlas-D as it came to be called, could be ready for deployment by the time the smaller warheads were available in the 1958-9 timeframe.

Also, the smaller missile would offer a major advantage in response time. The American missiles of the day, and presumably also the Russian ones, used liquid fuels. They could only be fuelled just before launch. Filling a 160ft tall, 12ft diameter Atlas with rocket propellant and liquid oxygen could take several hours. The smaller missile would significantly cut that time. (As it later turned out, the deployed Atlas-D had about a 15-minute launch preparation time [Parsch].)

Meanwhile, the assembly and testing of the Atlas-A (the original, heavy-lift version), proceeded as had been planned. It would be able to deliver in 1955 the heavy atomic warheads that were all the AEC was expected to be able to provide. If all went according to plan, this 'heavy-lift' Atlas-A would be operational perhaps for only a few years before being replaced by its smaller variant with longer range and shorter launch preparation. But those years would give the US an initial ICBM capability, and would minimise the potential 'missile gap' with the Russians. The Air Force justified the continued cost of the huge (and hugely expensive) Atlas A as being needed to prevent the Soviets from building up the lead in ICBMs which they might otherwise have for several years. Just before the 1952 presidential elections, the Air Force made their case to Congress for two versions of the Atlas. They did not mention the highly-classified additional mission they now planned for the Atlas-A.

As that election campaign progressed, the Atlas was barely mentioned. President Truman was still constitutionally eligible for another full term, but he chose not to run (Dallek). Among the issues bedevilling him in 1952 was the recent victory late the previous year of Mao Zedong in China (as discussed in Chapter 16).[8] Although Chiang Kai-shek's corrupt and incompetent Nationalists were largely responsible for their own defeat, Truman and his administration were catching much of the blame for the loss of China to the Communists. As a result of Truman's choice, and almost inconceivable from today's perspective, the 1952 contest involved two somewhat reluctant candidates: war hero Dwight Eisenhower on the Republican side and Illinois Governor Adlai Stevenson on the Democrats'. Each had to be persuaded to run by their respective party leaders (Parker). The election issues included government corruption and wasteful spending,

---

8. See 'What Really Happened' (1 Oct 1949).

inflation, and, of course, fighting Communism. In light of the all-but-certain 'loss' of China, Eisenhower promised a full review of America's military posture in Asia and worldwide.

Since space exploration was a popular topic at the time, it might have been expected to come up during the campaign. Willy Ley, a German-born scientist and popular writer, had published *Conquest of Space* in 1949 (Hardesty and Eisman), and other scientists and writers such as Fred Whipple and Arthur C. Clarke also helped stimulate a space craze. Clarke for example, predicted there would be a manned landing on the moon by 1980 (Hardesty and Eisman). An extended series of articles in *Collier's Magazine* starting in early 1952 played to and helped increase public interest in space. Under the banner *Man Will Conquer Space Soon*, prominent scientists such as Ley and Whipple described rocket ships, space stations and moon bases, even trips to Mars (Ryan). Several years later, entertainer Walt Disney approached Ley about appearing on his widely-viewed TV programme (Heppenheimer). Some older readers may remember the avuncular Ley with his German accent describing spaceflight to a national TV audience as part of a *Man In Space* episode of *Disneyland* (Ley).[9]

Nevertheless, the presidential election was a down-to-earth affair with hardly a mention of space issues. After 20 years with Democrats in the White House, the election was largely, as Eisenhower said, 'a time for a change' (Parker). He won easily, with Richard Nixon as his Vice President. As he had promised during the campaign, Ike did a thorough review of government spending especially for defence. He called it his 'New Look' (Heppenheimer). As a highly popular war hero, he could afford to be critical of wastefulness and duplication in Pentagon spending. Ike brought in businessman Charles Wilson from General Motors as his Secretary of Defense. Together, they sought to ensure that the Pentagon spent taxpayers' money as efficiently as possible, providing fully adequate national security without unnecessary expense.

This meant another round of bombers vs. missiles. Wilson, more familiar with tank engines than rocket engines, was sceptical of ICBMs (Sheehan). 'LeMay & Company' sought to exploit this in lobbying for more B-47s and B-52s to add to the existing fleet of B-29s, B-36s and B-50s. But not all Air Force leaders were intent on preparing to refight the previous war. Air Force Vice Chief of Staff General Thomas White wrote that he 'believed some airmen to be as wedded to the airplane

---

9. See 'What Really Happened' (9 Mar 1955).

as cavalrymen were to the horse', and 'the senior Air Force officer's dedication to the airplane is deeply ingrained, and rightly so, but we must never permit this to result in a "battleship" attitude [referring to some Navy officers' pre-war belief that battleships were the only major naval weapon that mattered]. We cannot afford to ignore the basic precept that all truths change with time' (Federation of American Scientists). He supported maintaining a robust, urgent ICBM programme.

These missile men found an ally in Vice President Nixon, who was especially concerned about the rise of Communist power and the need to counter it decisively. He believed deploying ICBMs was a top defence priority (Sheehan). How influential Nixon was in persuading Eisenhower is not known. Ike's military career had seen horse cavalry replaced by the primitive clunky tanks of the First World War and then the potent battle tanks of the recent war. He'd seen flimsy biplanes dropping bombs from the pilot's hand supplanted by B-29s dropping atom bombs. Ike understood the need to embrace change. So other Defense Department components took cuts, but the ICBM programme proceeded at full pace (Sheehan).

Accordingly, the heavy-lift Atlas-A advanced to a series of test launches from Florida's Cape Canaveral beginning in early 1954. There were several instances of what the rocket engineers began calling a 'RUD', or 'Rapid Unscheduled Disassembly' (Svitak) – they blew up. The occasional fireballs startled the native alligators but otherwise attracted relatively little attention in the sparsely-populated Florida marshlands. As General Schriever commented, 'If we were certain these contraptions would work, we wouldn't have to test them'. The Air Force analysed the problems of each launch attempt and continued the testing in relative obscurity throughout the spring of 1954. Then came the announcement from Moscow, and things changed dramatically.

Chapter 23

Soviet Rocketry after Stalin

Right after Stalin's death in 1953 no one knew who was in charge. For many months Khrushchev, Beria, Molotov, Malenkov, Bulganin and others jockeyed for position and power in the byzantine web of Communist Party and Soviet Government structures. In those early months rocket chief Sergei Korolev hosted a visit to his research facility by several of the Soviet senior officials, including Khrushchev. Korolev showed off a mock-up of the enormous R-7 rocket he had under development and he discussed its capabilities.[1] Many of the visiting government officials were lukewarm to this new-fangled technology, believing that the Soviet Union's main strength for the future would be in the ground and air forces that had won the Second World War. (As of the early 1950s the Red Army still had 260 divisions at a time when the US had just 10 active Army divisions [Heppenheimer; Coker].) But Comrade Khrushchev was visibly fascinated by the 110ft tall R7 rocket (Brzezinski). Like Stalin before him, he saw that the USSR needed to show its ability to project its power not just on the ground in Europe again, but into the American homeland. The more potent a threat the USSR could pose with its rockets, Khrushchev believed, the less it would have to spend on bombers and tanks and ground troops, and the more resources could go toward improving the daily lives of Soviet citizens (Cadbury).

Korolev of course took the lion's share of the credit for the R-7, although he made passing reference to 'the Germans' as having been 'somewhat helpful'. He knew that Khrushchev would be gratified that these Germans were making some reparation for the devastation the Nazis had wrought in Ukraine, where Khrushchev had served during the war. Korolev also used this opportunity to point out that the R-7 was

---

1. See 'What Really Happened' (Mar 1957).

not only powerful enough to carry an atom bomb into America, but was also capable of putting a satellite in orbit (Brzezinski). He described (von Braun's) concept of a series of satellites armed with atomic missiles. As von Braun is reported to have written: 'The first power that builds and occupies a space satellite will hold the ultimate military power over the Earth' (Cadbury). And 'No nation will challenge the Power that looks down upon it from such an artificial moon' (Ward).

Khrushchev was intrigued. He saw great propaganda value in being the first nation to put a satellite into orbit. Soon thereafter, he authorised Korolov to develop a satellite, but only as long as doing so did not distract him and his team from getting the R-7 fielded as an ICBM (Brzezinski). A satellite development team involving several academic and military institutions soon proposed a series of instruments that would fill the R-7 nosecone many times over (Heppenheimer). But the satellite team had only a matter of months to decide what to include and then to fabricate compact, spaceworthy equipment. The complex, rushed committee effort concerned Korolev; so he tasked another, smaller group to craft a simpler, smaller, backup satellite to be ready as soon as the R-7 was. Korolev was convinced that he was in a race against the Americans to launch the first satellite (Brzezinski). One of his key information sources was that 1952 series of space exploration articles in *Collier's Weekly* (Cadbury).

As to the missile itself, there were several failed tests of R-7 components in 1953. According to Russian accounts these were the inevitable result of design errors, with unanticipated and unforeseeable factors coming into play for the first time in such a huge machine. According to German accounts, the design flaws were typically undetected because interpersonal rivalries among the Russians sometimes took precedence over careful engineering and German advice. Other failures came about through complacency that allowed sloppy fabrication or assembly (such as valves installed backwards) at one or another of the Russian component factories (Heppenheimer). There was, however, extensive data on each failure. It was von Braun's recommendation, based on his wartime custom, to ensure that every test was thoroughly instrumented. Multiple sensors of various types would send back real-time information to the controllers on every aspect of the performance of each component (Sheehan). The Germans' passion for perfection could not prevent Soviet errors, but at least there was enough data to find and fix the problems.

By early 1954, Korolev and company decamped to the Tyuratam range in Kazakhstan to put the R-7 to the test. The first several resulted in what the Americans would call 'RUDs'. Either on the pad or very near it, they blew up. But after several failures, an R-7 flew 4,000 miles, with a 10,800lb payload (Sheehan). It was April 1954. In a sense, the ICBM era had started, although only the Soviets knew it.[2] The rocket had flown several hundred miles above the atmosphere into space (Wade), but the US had not detected it because they had no radars powerful enough or close enough to detect such a flight. However, the test was not wholly successful: the nosecone burned up on re-entry from space. The payload was destroyed high in the atmosphere (Brzezinski). Korolev was sure this problem could be solved, but until it was, the R-7 was useless for delivering a warhead.

Meantime, he decided to make lemonade out of lemons: he'd put the huge missile to the other use he and von Braun had long planned. He made the case to Moscow that orbiting a satellite which the Americans could detect would let them know that the USSR had an ICBM. Korolev assumed, correctly, that the Americans did not know about the nosecone problem. With Moscow's approval, he readied another R-7, with a satellite as payload.

As to the satellite to carry into space, the interagency team set up by the Kremlin had not finished their work on what was called 'Object D' (Zak undated [1]). The various instruments slated to go into space were not yet ready (Harford). Fortunately, Korolev had his alternative satellite ready. It was a 2ft polished metal sphere with four trailing antennas. It weighed 184lbs, most of it batteries for its radio transmitter. It would do little else but announce its presence by radio signal; perhaps its shiny skin would reflect enough sunlight to make it visible from the earth's surface (Zak undated [1]).

So instead of the heavily instrumented 3,000lb satellite as originally planned, an R-7 lifted off from Tyuratam and inserted into orbit a 184lb sphere as the world's first artificial satellite (Cadbury). It was 27 June 1954.[3] Called by the Soviets simply Sputnik (satellite) it announced its presence in orbit by sending out radio beeps heard by receivers around the world as often as the miniature new moon came within range.

Korolev may have been frustrated that Sputnik was such a small payload, when he knew that the R-7 could easily have inserted the far heavier 'Object D' into orbit. The R-7 was designed to develop over one million pounds of thrust (Wade) and could have delivered an 11,000lb warhead 5,000 miles away, or a 3,000lb satellite into orbit (Heppenheimer). Using all that power to only put 184lbs into orbit was akin to swatting a fly

---

2. See 'What Really Happened' (Aug 1957).
3. See 'What Really Happened' (4 Oct 1957).

with a hammer. Still, he (and the Soviet Union) had beaten the Americans into orbit.

Korolev's military employers were pleased too: the satellite showed the Americans that the USSR had a vehicle capable of delivering an atom bomb to the United States. The fact that the rocket men did not yet know how to prevent such a warhead from uselessly burning up in the atmosphere was not widely shared outside of Tyuratam. Sputnik was nonetheless generally a stunning achievement. Among the not-so-stunned were members of the Atlas team in the US. In fact, a few of the more secretive types were quite pleased at the news.

Chapter 24

Response to Sputnik

For at least seven months before it did happen, there had been reports from the intelligence community about a possible Soviet satellite launch within a year (Ryan and Keeley), so the news from Moscow about Sputnik was not surprising to the Pentagon. What was surprising to the Atlas team was the satellite's small size. 'If that's all they can put into orbit, it means they can't drop much on our heads, yet', observed Schriever.

Officially, the Eisenhower White House was nonchalant, even dismissive about one small ball in the sky (Ward). Ike called it a 'neat technical trick' and 'a feat of primarily Nazi engineering, not Soviet know-how'. It was 'a propaganda stunt, without military significance' (Brzezinski). Democrat Senator J. William Fulbright of the Senate Foreign Relations Committee also pooh-poohed Sputnik, saying 'it did not feed a single Russian. It was a trick, a gambit. So far as real prestige goes, it is nothing' (Mieczkowski).

If the Pentagon was not surprised to hear about, and to hear, Sputnik, the public and the media were. They did not like this surprise. Newspaper editorials quickly claimed that the Soviets' capability to launch a satellite indicated their capability to hit the US with an ICBM (Ward). Manhattan Project scientist Edward Teller, a better scientist than a military analyst, termed Sputnik 'a defeat more important than Pearl Harbor' (Hardesty and Eisman).

The Washington Post headlined 'Expert Sees Moon As Rocket Base'. It said,

> The Air Force's top space expert predicted yesterday the moon will be a military rocket base for either Russia or the US within 10 years. General H. Boushey, the deputy director of Air Force Research and Development said the moon will provide a 'base of unequaled advantage' for raining 'sure and massive destruction'

on earth. The General said he fully supports the view that 'He who controls the moon, controls the Earth'. (Wasser)

A few weeks later the widely-respected foreign affairs expert C. L Sulzberger wrote an analysis of General Boushey's views in *The New York Times*. It concluded, 'Such concepts are fantastic but no longer fanciful. And their potential military implication is immense. Manned platforms in outer space or missile ramps upon the moon would give the controlling nation a seemingly overwhelming advantage from which to dictate' (Wasser).

In Congress, the Democrats chortled privately while decrying the Soviets' achievement. Senate Minority Leader Lyndon Johnson, urged by his political advisor George Reedy to 'plunge heavily' into the issue, called the launch of Sputnik 'a latter day Pearl Harbor' (Wasser). Demonstrating that his strength was clearly greater in rhetoric than in science, Johnson also said 'Control of space means control of the world. From space the masters of infinity would have the power to control the earth's weather, cause drought and flood, change the tides and raise the levels of the sea, to divert the gulf stream and change temperate climates to frigid' (Heppenheimer).

Johnson continued: 'In essence, the Soviet Union has appraised control of space as a goal of such consequence that achievement of such control has been made a first aim of national policy. [In contrast], our decisions, more often than not, have been made within the framework of the Government's annual budget' (Wasser).

Bernard Baruch, statesman and advisor to FDR and to Truman, lamented 'While we devote our industrial and technological might to producing new model automobiles and gadgets, the Soviet Union is conquering space' (Brzezinski). He called for abandoning the 'post-war consumerism and its moral and spiritual flatulence', whatever that meant. Carl Sandburg blamed America's embarrassment on philistinism, whatever *that* was. Democrats such as Johnson and Stuart Symington criticised the White House for cutting costs and thereby jeopardising national security (Heppenheimer). The Soviet press, of course, extolled 'the advantages of the labour of socialist society in turning dreams into reality' (Heppenheimer). The Nazi contribution was not mentioned.

While the President was publicly trying to tamp down the public brouhaha, within the Administration he had to rein in his Vice President's political aggressiveness. Richard Nixon, likely even in 1954 eyeing his own eventual run for the White House, saw Sputnik as 'a challenge to

America's scientific and engineering ability to stay abreast of what was going on in the world'. Nixon feared that a flaccid response to Sputnik would hurt him politically in 1956 or 1960. He wanted a vigorous reaction and was frustrated by Ike's tepid response to Sputnik. At a White House meeting shortly after Sputnik went up, Arthur Larson of the US Information Agency proposed, 'an all-out project aimed at either hitting the moon, launching manned satellites or establishing a space platform'. The President was not interested, but Nixon was. The Vice President opined 'we can make no greater mistake than to see this as just a Soviet stunt. We've got to pull up our socks and get with it and make sure that we maintain our leadership' (Mieczkowski).

While Nixon fumed, Ike was asking the Pentagon when the secret American satellite he'd been briefed on in his first days as President would be ready. That satellite had its roots at the very end of the Second World War. As noted earlier, numerous research and development projects were funded in 1945 with leftover monies, and among these were studies on artificial satellites. After the Navy had approached the Army right after the war to collaborate on developing a weather reconnaissance satellite (RAND undated), the Army Air Force asked Douglas Aircraft to look into the feasibility of developing an earth-circling spacecraft (RAND undated). Douglas's research arm, Project RAND, reported that building a space satellite was clearly doable with the available technology. The RAND team believed that within about five or six years a multi-stage rocket derived from a scaled-up version of the V-2 could put a 500lb satellite into orbit (RAND 1946). The report concluded that such a satellite could return intact to Earth after about 10 days, and further stated that successors to this initial satellite could very well be manned. A satellite, the report envisioned, could serve to provide target guidance to long-range missiles, or could perhaps serve as an orbiting weapon itself, ready to strike an earthly target on command. The report noted also that a rocket that could launch a satellite would be very similar to one that could be used as an ICBM (RAND 1946).

The RAND team was of the strong opinion that

> a satellite vehicle with appropriate instrumentation can be expected to be one of the most potent scientific tools of the twentieth century and that achievement of a satellite craft by the United States would inflame the imagination of mankind and would probably produce repercussions in the world comparable to the explosion of the atomic bomb. We can see no more clearly all the utility and implications of spaceships than the

Wright brothers could see fleets of B-29s bombing Japan and air transports circling the globe (RAND 1946).

One of the RAND report's principal authors soon after took a different tack in expressing the value of developing a satellite:

> Since mastery of the elements is a reliable index of material progress, the nation which first makes significant achievements in space travel will be acknowledged as the world leader in both military and scientific techniques. To visualize the impact on the world, one can imagine the consternation and admiration that would be felt here if the United States were to discover suddenly that some other nation had already put up a successful satellite (Campbell).

The Air Force, facing tightening budgets, chose not to actively pursue the satellite programme in 1946. But they tasked Project RAND (soon to spin off from Douglas as the separate RAND Corporation) to continue studying the satellite idea. This they did, submitting at least half a dozen classified reports to the Air Force from the late 1940s through the early 1950s (Lipp and Salter) with titles such as 1951's *Utility of a Satellite Vehicle for Reconnaissance* (Heppenheimer). In a report on Project Feedback, RAND analysts concluded that reconnaissance satellites with TV cameras could overfly the Soviet Union and scan the landscape for airbases and missile facilities (Lipp and Salter). They estimated that such satellites would weigh about 4,500lbs and could be lifted into orbit by a rocket with the power of the ('heavy lift') Atlas then under development (Lipp and Salter). With the Atlas slated to fly in the 1954 timeframe, the RAND report indicated that a reconnaissance satellite could be ready at the same time.

However, the authors predicted a potential geopolitical brouhaha if the US were found to be violating the Soviets' sovereign territory with a spy in the sky over Mother Russia. At the time, the secretive Soviet Union, unlike most other nations, would not even allow foreign commercial aircraft to overfly its territory (Heppenheimer). So the RAND authors urged that the project be kept secret as much and as long as possible, using if necessary a cover story of some sort to mask its espionage mission.

Accordingly Project Feedback was never heard about again. Anyone who might have probed the subject would have found that RAND's TV idea wouldn't work: researchers soon found that the concept of transmitting the satellite's recorded TV images back down to earth whenever the satellite came within range of a US ground station was unworkable (Ruffner). This was ostensibly why the project died.

But it didn't; it just went dark.

In its new configuration, the planned reconnaissance satellite would use a powerful new high-resolution film camera that could take detailed photos of the earth's surface from space. This 'HYAC' camera had been developed for the GRANDSON programme, a 1951 Air Force effort to use high-altitude observation balloons (Welzenbach). That Air Force plan envisioned releasing huge balloons over northern Europe with cameras in their gondolas. The balloons would float about 20 miles high and cross the Soviet landmass on the prevailing winds. Out over the Pacific several days later, the balloon would release the gondola, which would descend by parachute. An Air Force cargo plane with a special net extending from its rear cargo door would catch the gondola and thereby retrieve the camera and its intelligence-laden film (Welzenbach). It took many attempts, but by 1954 the Air Force had developed improved techniques for plucking the gondola, the equivalent of a falling refrigerator, from the sky (Welzenbach). Although this somewhat wacky-sounding retrieval technique worked, the overall scheme did not – the film contained too many images of clouds, and the Russians kept protesting about the errant 'weather balloons' wafting over their territory (Wasser). Project GRANDSON was abandoned.

But the airborne retrieval technique became available just in time to solve the retrieval problem for the spy satellite under development. Project Feedback had evolved into a deeply classified effort later known as Corona (Ruffner). (Despite latter-day attempts to explain the name as a reference to the sun's plasma envelope, the more prosaic truth is that a CIA official took the name from the typewriter on his desk, a Smith-Corona [Heppenheimer]. Nor, despite would-be conspiracy junkies, does this name have anything to do with coronaviruses, which were first identified in the 1960s [Lim et al.].)

As the project took shape, the plan called for the spy satellite to ride into space on a (heavy lift) Atlas. Its cameras would take about three miles of film as the satellite passed over Russia (Drell). Then it would jettison the film, which would re-enter the atmosphere in a heat-shielded capsule. The capsule would then parachute down toward the ocean, and an Air Force plane would catch it before it hit the water (Ruffner).

Thus, when Sputnik went up in mid-1954, and Ike asked about the US satellite programme, the Air Force and CIA team told him that the

Corona satellite was ready. Despite the national embarrassment, there was an advantage to the US in the Soviets getting into orbit first. For some time, Air Force, RAND, State Department and other analysts had grappled with the international legal implications of satellite overflights (Oder et al.). Did a nation's sovereignty extend into all the space above its territory (*ad coelom* – to the heavens)? Or was there a definable upper limit, above which there was a 'freedom of the seas' as in maritime law (Hardesty and Eisman)? It was not just an academic discussion. The US very much wanted to see what was going on in Russia especially in regard to their bomber and missile capabilities.

At the time of Sputnik, Eisenhower was beginning to formulate a proposal he planned to make to the world community. His 'Open Skies' concept would have explicitly allowed the US and USSR to conduct aerial reconnaissance overflights over each other's homelands (Oder et al). Now in mid-1954 the arrival of a Soviet satellite overflying US territory every day obviated the lawyers' and policymakers' concerns about the legality of a US satellite doing the same to the Soviets. 'The Russians probably wouldn't have agreed to that Open Skies idea anyway', Ike commented to his advisors.[1] In that sense, Sputnik won prestige for the Soviets, but lost them security by freeing the US from its caution about using spies in the sky (Heppenheimer). By being the first to launch a satellite, and having it fly over the United States, the Soviets had lost their ability to object diplomatically. Ike was secretly well pleased (Wasser).

The US (spy) satellite was ready, but its ride into space, the Atlas, was not quite there. As noted, the early 1954 test launches of the Atlas had not been successful. The US Navy, ever ready to lend a hand to a sister service, promptly came forward with a proposal to use its Viking sounding rocket to go ahead with launch of a scientific research satellite (Green and Lomask).The Navy had been using the Viking, derived in large part from the V-2 design, for upper atmosphere research since the late 1940s (Rosen). Vikings had flown as high as 150 miles, sending back data on the atmosphere, and on cosmic rays, as well as high-altitude photos of earth (Rosen). The Navy had a plan to use the Viking as a first stage, and its Aerobee sounding rocket as an upper stage. The Aerobee too had been in use for atmospheric research for several years (Van Allen), but the two rocket systems had never been used together. The Navy projected that this lashed-together system could insert a science satellite of several pounds into orbit, and it would make

---

1. See 'What Really Happened' (Jul 1955).

observations on characteristics of the space environment. The Naval Research Lab had been working on this idea for some time and they told the White House they could launch within two or three months. Ike authorised the Navy to proceed with preparations to launch its Project Vanguard, but it would only be as a backup if the Air Force was unable to launch the Corona with its Atlas.[2]

Nor would the US fully disclose at the time what the Corona satellite was for. Instead, as the RAND report had recommended years before, the spy satellite got an elaborate cover. What the Air Force would announce was that the satellite was called Discoverer, an instrument to 'gather much needed scientific information on the earth's atmosphere out to 125 miles' (Oder et al.). Subsequent satellites in the programme, the Air Force said, would also include biomedical observations with mice and primates, which would be retrieved after their ride into space (Ruffner). What the Air Force did not say was that along with the instruments for measuring such things as atmospheric density and heat transfer, and eventually the condition of the spacefaring lab animals, the satellite would also have a pair of powerful cameras and miles of film, busily taking photos of the goings on in the USSR. So safely retrieving the critters in the capsule would not be nearly as important as the film canisters in the hidden compartment as the whole re-entry vehicle fell out of the sky (Ruffner).

In August 1954 an Atlas-A (the heavy-lift version) flew successfully from Cape Canaveral to its full 5,000+ miles range, far into the South Atlantic.[3] The President then authorised the next launch to carry a Discoverer satellite. Accordingly, on 2 September, 1954 an Atlas-A launched America's first satellite.[4] Discoverer 1's orbit was intentionally similar to that of the Sputnik, which was circling the earth about every 90 minutes at an angle of 60 degrees from the equator. Just as Sputnik was flying over the US multiple times each day, Discoverer now repeatedly flew over the Soviet landmass daily.

Also like Sputnik, Discoverer broadcast a beeping radio signal which people around the world could hear. But unlike the 180lb Sputnik, Discoverer was said to be rather more massive, about 1,500lbs. It was also said to be steadily radioing back a stream of data from its several instruments.

---

2. See 'What Really Happened' (3 Dec 1957).
3. See 'What Really Happened' (Jun 1957).
4. See 'What Really Happened' (31 Jan 1958).

> **CLOAKING DEVICE**
>
> The shielding of the Corona spy mission with the Discoverer science programme was thorough. Many people working on the 'biomedical science package' were never told what the real mission was. Although much has been released about the Corona/ Discoverer programme in recent years, some aspects are still classified more than half a century later (Oder et al.).

The two months of America's embarrassment had come to an end, with a clear demonstration that the US had the capability not just to launch satellites, but large, sophisticated ones at that (even though only a handful of people really knew how sophisticated the Discoverer/ Corona vehicle really was). The public reaction was, if not jubilant, at least one of great relief that although the Soviets had gotten into space first, they had only sent up a comparative gimmick.

In the Soviet Union, Korolev was reported to have said unflattering things about the academics who were still puttering around with 'Object D'. Had it been ready in June, its launch would have been a far more impressive blow to America than the little Sputnik was. Even in September of 1954, the complex package of instruments was still not ready to fly, so Korolev improvised a second satellite to more clearly demonstrate Soviet capabilities. Russian sounding rockets had been carrying dogs as high as 60 miles into space since 1951 (Heppenheimer). Korolev now fitted one of these dog-pods, and accompanying instruments, into an R-7 nosecone and launched it (Heppenheimer). On 19 September 1954 the Soviets thus put the first living creature into orbit. They demonstrated that living organisms can survive the stresses of launch, and the subsequent conditions of weightlessness. Sadly, the dog died after some time in orbit due to oxygen exhaustion, the Soviets said. The world was generally impressed with the dog; US military observers were impressed with the 1,100lb weight of the capsule. Sputnik 2 showed that the Soviets had the capability after all to deliver nuclear warheads at intercontinental distances, as did the US.

America's Discoverer 2 went into orbit soon thereafter, in October 1954. It was the first to carry a camera, but that part was not announced. What was touted was its cargo of several mice, who returned safely to Earth a few days later. This live recovery shifted the bragging rights back to the US. The successful mission was somewhat of a pleasant surprise for many of the programme personnel, who had believed that it might take a number of attempts to master the series of intricate steps involved. Years later one manager remarked, 'The enormous power of that Atlas booster let us build and launch a robust system with lots of redundancies – like extra parachutes and other measures. If we had been more constrained by weight, it might have taken many tries before we got a successful launch and recovery.'[5]

What was not talked about at the time was that the same capsule that carried the bewildered (but still healthy) mice also contained film that yielded thousands of photos of the Soviet Union. There had been an elaborate shell game-like procedure, orchestrated by the Air Force, whereby officials posed for photos with the recovered science package before sending it to the lab, while the film canisters formerly attached to the mouse pod, and which officially did not exist, travelled through black channels to CIA analysts in Washington (Oder et al.). The existence of the resulting photos was known to very few, but when Ike saw a few of the best ones, he was highly pleased. The camera had photographed 1.5 million square miles of the Soviet Union and Eastern Europe (equivalent to half the area of the lower forty-eight states), from which sixty-four Soviet airfields and twenty-six new anti-aircraft missile sites were identified (Ruffner). Ike now knew that the US had a way to 'peek through the keyhole into the Soviet house and see for ourselves what was going on'. (Later US spy satellites adopted this image with their 'KH' prefix.) There were many more Coronas launched over the years, some with two cameras and two recoverable capsules. Other satellites also included devices that worked with other than visible light (Oder et al.) The CIA has called Corona's contribution to the intelligence effort 'virtually immeasurable' (Oder et al.)

Ike also realised that the much riskier project he had recently approved could be re-evaluated. In November 1954 the CIA was set to contract with Lockheed Aircraft for a very fast-track development of an exotic high-altitude spy plane (Pedlow and Welzenbach). It was designed to fly higher than Soviet radars could detect, or, failing that, higher than Soviet anti-aircraft missiles or fighters could shoot down. Essentially, it was intended to fly and spy in Soviet sovereign airspace, an act that was, if not an act of war, certainly not a peaceful one. Ike had been willing

---

5. See 'What Really Happened' (Aug 1959).

to risk the potential outcry from the Kremlin if the plane could obtain useful photos of what was happening on the ground. Now, with a space-based tool, the urgency of the manned spy plane programme, if not its entire utility, decreased.[6]

Therefore, the Pentagon put the rushed procurement on hold while they re-evaluated the need for the spy plane. After some months, the CIA and the Department of Defense agreed that the flexibility and responsiveness of a manned aircraft would complement the capabilities of orbiting spy platforms. 'Once in orbit, a satellite has limits on what it can fly over and when. A plane can go where we want, when we want', was the reasoning. So, with White House approval the aircraft later dubbed the U-2 did get the go-ahead in early 1955, but on a much less urgent, 'more cost-efficient' basis than originally planned. It first flew four years later, with disappointing results. Its design had been 'improved' over the years by the multiple review panels that got involved. The result was a more robust (heavier) aircraft that could not fly as high, as fast or as far as the original design. Lockheed's project manager, Kelly Johnson, reflected years later, 'If they had let us build the U-2 as first planned, we could have flown a better aircraft in less than a year.' Few observers believed his boast, except those who knew him and the Lockheed 'Skunk Works' team.

For their part, the Soviets threw yet more weight into orbit in 1955, finally launching their 3,000lb 'Object D' as Sputnik 3.[7] The US continued launching Discoverer (Corona) satellites, ostensibly sending back instrumented science packages. The story was that they contained various instruments that needed to be visually inspected to determine the effects of spaceflight (Oder et al.). *The New York Times* praised the capabilities of the Pentagon in recovering these science packages: 'The feat marks an important step toward the development of reconnaissance satellites that will be able to spy from space. The same ejection and recovery techniques eventually will be used for returning photographs taken by reconnaissance satellites' (Ruffner). Presumably *The Times* did not realise that their 'eventually' was 'already'. It is not known how long it took the Soviets to figure out that the Discoverer satellites were just the cover for spy satellites. At any rate, it was not long before the Soviets were returning the favour, launching their Zenit series of photoreconnaissance satellites with film recovery under the ostensible

---

6. See 'What Really Happened' (Nov 1954 and 1 May 1960).
7. See 'What Really Happened' (May 1958).

cover of the Kosmos series of scientific exploration satellites (National Air and Space Museum).

While the public followed the space exploits of the superpowers, their militaries focused on the weapon systems that had given rise to those exploits.

Chapter 25

Missiles East and West

Khrushchev was pleased that the early Sputniks were a propaganda win for the Soviets. Yet he knew, as his generals did, that the military significance of the new rockets was potentially far greater. He boasted shortly after Sputnik that the USSR was 'turning out missiles like sausages' (Sheehan). When his son privately questioned the accuracy of that statement, Khrushchev replied, 'It's alright; we don't have so many sausages either'. But they did have the R-7, and the Soviets focused their production on deploying as many as they could, knowing that the US was not far behind with their Atlas. The primary value of the R-7 was that it existed. It demonstrated technological capability. It posed a theoretical threat of swift, unstoppable nuclear attack against the US. However, it was not really militarily useful, since it would take almost 24 hours to prepare it, fuel it and launch it, hours during which it would be vulnerable to US bombers or missiles (Brzezinski).

Korolev and his Soviet (and German) colleagues were already working on the R-9, a more accurate ICBM capable of launching on about 20 minutes' notice (Wade). Their success in fielding the R-7 had helped them fend off earlier Kremlin calls to develop shorter-range missiles as an interim expedient. With such 'medium and intermediate range ballistic missiles' Russia could have threatened European members of NATO. Or, as one of the rocket men put it, 'If we had medium range missiles, well, then we could threaten the US mainland – as soon as the Kremlin gets us basing rights in Mexico or some place in the Caribbean'. The (German-aided) success of the R-7, however, meant that the Soviets need not expend resources on shorter-range missiles, or on trying to get them based in the Caribbean. They could concentrate on fielding homeland-based ICBMs to counter the Americans' ICBMs, which of course were being fielded to counter the Soviets' ICBMs.[1]

---

1. See 'What Really Happened' (Aug 1955).

For their part, the European NATO members had no need for medium or intermediate-range missiles on their territories, as long as Russia and the US were positioning themselves for a strictly long-distance confrontation. There was no need for the potential geopolitical complications that would have arisen if, for example, the US had developed nuclear-armed intermediate range missiles and placed them in Turkey, Italy or the UK.[2]

During 1955–6 the Corona spies in the sky kept watch on the Soviets' new launch facilities and associated production and transportation capabilities. By the end of 1956 the Kremlin had deployed twenty-four R-7s, and had launch facilities under construction for several dozen more. The R-9s were reportedly slated to begin deployment in 1957. It wasn't quite a matter of 'turning out missiles like sausages'. Still, the available data did indicate that the Soviets, by concentrating their efforts on the R-7, and then the R-9 ICBMs, would perhaps have several hundred ICBMs by 1959. The visible pace at which the Soviets were putting a sizable ICBM force in place was of concern to US analysts. The data suggested that by 1959 the Soviets would have several times as many missiles fielded as the US, with the possibility that such a ratio would persist. Ike was pleased to have such good intelligence, but not happy about what the photos and other intelligence indicated.

As to bombers, Ike knew in 1955 from Corona and other sources that there was a huge bomber gap: the Soviets were far behind the US and not likely to catch up. The first 20 of a planned 700 intercontinental-range B-52 jet bombers were being fielded by the USAF that year. They would join the existing SAC fleet of 250 B-36s and 1,300 B-47s (National Resources Defence Council). Against this air armada Ike knew that the Soviets had a few hundred Tu-4s, their copy of the old US propeller-driven B-29. These did not have the range to reach the US. The Soviets' M-4 'Bison' jet bombers perhaps could reach the US, on a one-way mission, but there were fewer than twenty of these.[3]

Ike was sure that the US ability to inflict nuclear devastation on the Soviet Union far exceeded anything they could do to the US. But he also knew, in this new era of rockets and space flight, that allowing the Soviets to hold a significant edge in missiles would be unacceptable, even though the US edge in bombers was overwhelming. It was not a matter of tonnage but time. Manned bombers would take many

---

2. See 'What Really Happened' (Nov 1955 and Oct 1963).
3. See 'What Really Happened' (July 1955–1958).

> **FLIM-FLAM FOILED**
>
> During the Moscow Air Show parade in 1955, the Soviets staged a flyover by two groups of its new 'Bison' long-range bomber, showing twenty-eight aircraft in all. The clear intent was to show domestic and foreign observers that the Soviet Air Force had a goodly number of these powerful jets. But there was less there than met the eye: the ten aircraft in the first flyover had looped around out of sight and joined the eight of the second flight (Sheehan). The ruse was convincing to observers on the ground, but CIA and USAF analysts, who counted just eighteen 'Bison' in Corona photos before and after the parade, advised the White House of the trick.
>
> Eisenhower, an inveterate poker player and world-class bluffer (E. Thomas), reportedly commented to an aide, 'Nice try, Nikita.'

hours to reach their targets in the USSR. Missiles would take about 30 minutes between the USSR and the US. That asymmetry was more than disquieting to the White House.

Accordingly, in early 1956 he directed the Air Force to accelerate its planned Atlas deployments. Then he reluctantly went to Congress for additional funds. The Democrats, especially those eyeing the White House themselves, such as Stuart Symington, Lyndon Johnson and John Kennedy, squawked a bit. 'We are already funding multiple kinds of bombers and long range missiles; now you want yet more missiles?' Without revealing Corona specifically, the White House explained the intelligence community's assessment, and asserted 'We must close the Missile Gap. We cannot allow our adversaries to think they can ever intimidate us with their missiles. They need to know that our Atlas missiles are available to ensure our national security.'

After the customary and predictable back and forth, Ike got additional funds to buy more Atlases, but not as much as he asked for. However, Congress also allowed the Pentagon to re-allocate some funds from bombers to missiles, because no one on Capitol Hill wanted to be accused of worsening the Missile Gap. USAF Chief Curtis LeMay griped about losing some bombers, but associates said he was grinning behind his ever-present cigar as he thought about the additional Atlases.

In January 1956 the first US ICBMs (six of the big Atlas-As) had become operational at Plattsburgh AFB in northern New York State.[4]

---

4. See 'What Really Happened' (Sep 1959).

Their northerly basing location put a large portion of European Russia within range. Like their Soviet R-7 counterparts, these massive missiles could not have been fuelled and launched rapidly, but their primary value was as a demonstration of capability to the Soviets, and to the American public. In the next 18 months, several dozen more were deployed at Fairchild AFB Washington, Malmstrom AFB Montana and Minot AFB North Dakota. These missiles closed the short-lived missile gap, albeit with a bit of bluff. About sixty of the first American nuclear-tipped missiles capable of hitting the USSR were declared 'operational', about a year before their crews were sufficiently well trained to be able to operate them. That is, a year after the US told the world these missiles were ready, they actually got their warheads (Sheehan).

At the same time, the smaller Atlas derivative, the Atlas-D, capable of delivering a smaller (2,000lb) payload almost anywhere in Russia, also proceeded through design, testing and then to deployment. This missile could be fuelled and launched much more rapidly, from hardened, blast-resistant concrete silos. This programme proceeded on Ike's accelerated schedule, deploying the first units in early 1958. These were scattered among farm and ranch lands around Air Force bases deep in the US interior, like Warren AFB Wyoming, Omaha's Offutt AFB, and Whiteman AFB in Missouri. As the USAF deployed more Atlas-Ds, they phased out the huge Atlas-As. But all of these found alternative uses as satellite launchers. Derivatives of Atlas, America's first ICBM, are still in use today, launching government and commercial satellites (Parsch; Davenport).

Ike closed the Missile Gap before it had become problematic. We cannot know if his decisive action in response to Soviet ICBM development affected his re-election in 1956. That contest was one of the rare rematches in Presidential races, in which the same two candidates faced off in successive elections. In the three previous times such a rematch occurred (1824 and 1828; 1836 and 1840; 1888 and 1892) the results changed, i.e. the winner the first time lost the second time. So, when the Democrats again nominated Adlai Stevenson to run against Eisenhower in 1956, they had some history on their side. Perhaps the pattern would have continued, and Stevenson would have won the rematch, if Eisenhower had not taken such a strong position against the Soviets in 1956, and prevented the Democrats from having a Missile Gap issue to use against him.

Chapter 26

The Continuing Space Race

The Soviets used dogs; the Americans used monkeys to make sure that higher animals really could survive the rigors of spaceflight. They could.¹ What they couldn't survive were the mechanical failures in their capsule life support systems, or their parachute recovery systems. Both sides thereby killed animals in their quest for space. As systems became more reliable, and confidence grew in regard to survivability, both nations took the logical step of putting humans into orbit. The Soviets went first, putting Yuri Gagarin into orbit on 4 October 1957 using a modified R-7.¹ Gagarin circled the Earth in a Vostok capsule, modified from the Zenit spy satellite (Gorin). By adapting these military assets for space exploration uses, Korolev satisfied the Defence Ministry that the latter was not detracting from the former. This was also a way to get the military to bear most of the cost of the civilian effort.

Gagarin returned safely to Earth and was feted worldwide. The international one-upmanship continued as the US put its first citizen into space on 31 January 1958. Alan Shepard rode a 2,700lb Mercury capsule atop an Atlas-A into orbit, and then circled the earth three times before splashing down in the Atlantic a few miles from the US Navy recovery ships.² Shepard soon thereafter rode parade routes from New York City to San Francisco. The new space agency NASA let it be known, with a bit of public relations exaggeration, that not only had Shepard taken multiple trips around the world (vs. Gagarin's one), but also that America's first 'astronaut' had actively flown his Freedom 7 spacecraft, and not just ridden in it, as Gagarin had done in his Vostok. NASA also claimed, just as the Kremlin had, that

---

1. See 'What Really Happened' (12 Apr 1961).
2. See 'What Really Happened' (5 May 1961).

their manned space flights and the data they obtained from them, were important contributions to the IGY, the International Geophysical Year of 1957–8 being sponsored by scientific organisations worldwide. A few people may even have believed that the space shots were primarily motivated by a spirit of international scientific cooperation.

Shepard's orbital flight was the first of the planned series of manned flights of Project Mercury. Its official objective was to 'investigate man's ability to function in space' through orbiting and then safely recovering a manned spacecraft (NASA). Although space travel had been a staple of stories and articles for many decades, it was actually far from certain that humans could survive and function in space. For several years, the US Air Force aeromedical programme had studied human factors in experimental high-flying planes, but the effect of sustained weightlessness in space on human performance was unknown. In the absence of earthly pressure on nerves and muscles, and with the labyrinth of the inner ear perhaps unable to signal the brain about which way was 'up', there was concern that humans would be too disoriented to perform required tasks. There was also concern about the effects of sharp acceleration and deceleration, and space radiation (Swenson et al.).

So it made scientific sense to say that investigating man's ability to function in space was a Mercury objective. But of course this prompted the question – why? Why was it worth spending huge amounts of taxpayers' money to see how well humans could work in space? The question was seldom asked, perhaps because there were several obvious answers. For one, it was necessary to know how well humans could work in space so that they could operate the orbiting observation platforms envisioned by the military (Swenson et al.).[3]

Such 'space stations' were also obvious stepping stones for the next logical step, a manned mission to the moon. In these first years of the space race, there was little doubt that putting humans on the moon was the overarching goal. Indeed, even in the months right after Sputnik, a USAF Man-In-Space Task Force had developed a seven-year proposed programme for a manned lunar landing (Swenson et al.). Within two years of Sputnik's launch, NASA scientists made a presentation at an International Astronautics Conference that depicted multiple spacecraft landing on the moon, and astronauts exploring the landscape (Brooks et al.).

---

3. See 'What Really Happened' (Dec 1963).

Of course, going to the moon had been a tantalising aspiration for many generations, especially since Jules Verne's 1865 *From the Earth to the Moon* described a technologically plausible way to do so (via a huge cannon located on the Florida coast near Cape Canaveral). Now that massive rockets capable of doing the job were within reach, there was no real question but that humans of the current generation would go to the moon. The questions that did arise concerned how a moon programme would proceed, on what schedule, and who would succeed first.

There was very little discussion of why. The Mercury manned spaceflight programme was expensive; the manned moon mission would dwarf it. This made for tension with the budget-minded Eisenhower White House, not so much as to whether there should be a moon mission, but as to how fast it should proceed (i.e. how much of its costs a future President and Congress would have to haggle over).

When, two years after Sputnik, the Luna 2 probe landed on the moon and deposited the Soviet flag on lunar soil (Bell), there was concern in the US that the Russians would seek to claim the moon as Soviet territory. The State Department told concerned Congressmen that 'there is considerably more to establishing sovereignty than planting a flag' and 'a large body of law already exists which could be expected to govern man in space just as it did on earth'. Presumably, this meant that the State Department believed the Soviets would actually have to put a man on the moon to claim it (Wasser).

Few if any Americans believed the US should attempt to claim the moon, but they did believe they should seek to pre-empt the Soviets from doing so. The newly-formed NASA held a planning conference at which one of the leading rocket developers expressed confidence that the US could develop, within 9 or 10 years, 'a capability of putting . . . man on the moon. And we still hope not to have to go through Russian Customs there' (Brooks et al.).

But the real impetus to go to the moon, and to do so as soon as feasible was the less tangible, but nonetheless real quest for 'national prestige'. Hanson W. Baldwin, the influential military affairs correspondent for the *New York Times*, had criticised the Eisenhower administration underestimating the prestige value of the Soviet's Sputnik. He chided the administration for neglecting the power of intangible ideas. He advised the government to seek more advice from political rather than physical scientists: 'It is not good enough to say that we have counted more free electrons in the ionosphere than the Russians have . . . we must achieve the obvious and the spectacular, as well as the erudite and the obscure' (Swenson et al.).

Eisenhower set up a Man-In-Space panel to advise him. It reported:

> We have been plunged into a race for the conquest of outer space ... As a reason for this undertaking some look to the new and exciting scientific discoveries which are certain to be made. Others feel the challenge to transport man beyond frontiers he scarcely dared dream about until now. But at present the most impelling reason for our effort has been the international political situation which demands that we demonstrate our technological capabilities if we are to maintain our position of leadership. For all of these reasons we have embarked on a complex and costly adventure (Brooks et al.).

Even before Alan Shepard's inaugural flight as Mercury's first astronaut in 1958, NASA had announced that Mercury would be followed by Project Gemini, involving two-person flights, and Project Apollo, which would involve three-person spaceflights. On a schedule to be determined, in a spacecraft not yet developed, launched on a booster not yet in existence, the initial Apollo flights would either be sustained orbital excursions or circumlunar missions, with the clear ultimate objective of a manned landing on the moon (Swenson et al.). As NASA explained: 'A primary reason for this choice was the fact that it represented a truly end objective which was self-justifying and did not have to be supported on the basis that it led to a subsequent more useful end' (Brooks et al.).

Ike approved NASA's initiating Project Apollo as a lunar mission, but without much specificity as to how it was to proceed, or how fast. As his Man-In-Space panel had pointed out, 'any of the routes to land a man on the moon will require a development much more ambitious than the present booster program, calling not only for larger boosters but for lunar landing and takeoff stages as well' (Brooks et al.).

Eisenhower supported US participation in the space race, but not at a pace that would impose excessive technical risk or strain on the Federal budget. Mercury still had to prove that spaceflights of more than a few hours were feasible. Because of the long lead times required, beginning development of a booster and spacecraft for Apollo was reasonable even before Mercury was finished, but this initial study phase did not cost very much or make an irreversible commitment to the programme.

There were even some attempts to manage public expectations about the outcome of the race. A Congressional report sought to prepare Americans for the possibility that a Russian might be first on the moon:

'Getting there first' is only one part of the race. Two other parts are just as crucial: 1. What will we learn from our effort to explore beyond the Earth? And 2. How will we use this knowledge after it is acquired? The Vikings had the technique to get to the New World 'first,' but England, France, and Spain won the prizes . . . The American effort has - thus far at least - been outstanding in its scientific results . . . Our international prestige and stature, so far as they are influenced by our space activities, depend on all three elements of 'the race' - not on one or two (Swenson et al.).

Alan Shepard's three-orbit flight in January 1958 was followed by a similar flight by Gus Grissom in April, and a nine-hour, six-orbit flight by John Glenn in July. Scott Carpenter and Wally Schirra flew even longer in 1959 and Gordon Cooper spent 34 hours in orbit in January 1960. As the 1960 presidential election campaign began taking shape, NASA's Project Gemini began launching two-man crews into earth orbit using Atlas rockets with added booster engines. The first of these was a four-hour flight by former USAF pilots Gus Grissom and Francis Gary Powers. But this did not occur before the Soviets had orbited Voskhod, the first multi-man spacecraft on 20 July 1960. Literally one-upping the Gemini flights, the Voskhod carried three cosmonauts. Like the Sputnik, this Soviet achievement may have been counterproductive. At the time, many US observers believed that the three-man mission indicated the Russians already had a robust spacecraft capable of carrying men all the way to the moon. (Only many years later did the Americans learn that the Voskhod capsule was really just a two-man earth-orbital vehicle into which a third seat had been crammed. Korolev and company had only been able to clear enough room for three cosmonauts by dispensing with their protective spacesuits [Heppenheimer]. It was not a vehicle that could have carried men to the moon and back.)

The prospect of what could be an imminent Russian manned moonshot was of great concern to many observers, including Richard Nixon. The Vice President had chafed for years under Ike's restraining influence, especially regarding the space race. As noted earlier, Nixon had lobbied hard early in the Eisenhower Administration for putting top priority on the development of the Atlas ICBM, which was now proving so useful as a space launcher. And later, right after Sputnik, Nixon had been frustrated by Ike's tepid response. Nixon, more so than Ike, had seen Sputnik as a threatening challenge to American technological and geopolitical leadership. The Vice President had not forgotten the suggestion by Art Lawson of the US Information Agency right after Sputnik that the US should undertake an all-out project aimed at the moon.

Nixon had no credible opponents for the presidential nomination at the Republican convention that opened on 25 July 1960. The Democrats had nominated Senator John F. Kennedy at their convention just before the Voskhod launch. In the Senate, Kennedy had not taken any particular interest in the space race, but his running mate, Senator Lyndon Johnson, had criticised the Eisenhower White House early and often about Sputnik ('a latter day Pearl Harbor'). As the Republican nominee and no longer just Ike's Vice President, Nixon now had the opportunity and the obligation to speak his own mind. Although he and Ike had disagreed on various issues during the previous eight years, Nixon knew that now he could count on the President to back him up, if not out of affection, then out of Ike's dislike for Kennedy. Ike had called Kennedy, 'that callow scion of a bootlegger's dynasty' (E. Thomas). Ike had also said, 'I'll do almost anything to avoid turning over my seat and the country to Kennedy' (E. Thomas). Polls before the conventions had suggested the race would be close. Nixon knew he needed every edge.

The editors of the magazine *Missiles and Rockets* had addressed an open letter to both the Republican and the Democratic Presidential candidates, inviting comments on a 'modest proposal for survival'. The journal sought specific commitments on the recognition as national policy of the strategic space race with Russia and on the endorsement of a bold long-range programme for space projects during the next decade (Swenson et al.)

Unknown to the magazine editors, Eisenhower had recently tasked his science advisor George Kistiakowsky to conduct a study of NASA's manned space programme. The study team were analysing the cost and the feasibility of sending astronauts around the moon by about 1970, and making a manned landing by 1975 (Heppenheimer). Nixon was familiar with Kistiakowsky's study. Nixon also knew that space was not a major issue or area of expertise for Kennedy. Hugh Sidey of *Time* magazine wrote some time later about Kennedy that of the various issues, 'he knew and understood least about space' (Heppenheimer).

This then was the background to Nixon's famous acceptance speech at the Republican convention on 28 July, in which he declared that the current Administration had put Americans in space, and that his Administration would send them to the moon. If elected President, he said, he would give the Apollo Program the specific goal of landing men on the moon and bringing them safely back before the end of the following Presidential term.[4] Chicago's International Amphitheatre shook. Ike, watching on TV, shook his head. The report he'd gotten from Kistiakowsky's panel envisioned a $40 billion price tag. Kennedy, watching Nixon's announcement on television, grimaced. Johnson cussed.

History tells us what happened that November. We cannot know how much Nixon's Moon Shot speech affected that outcome. Nor can we know how different the modern world would be if that election had come out differently.

---

4. See 'What Really Happened' (25 May 1961).

Chapter 27

Alternative Speculation

We have discussed how major aspects of the modern world were shaped by the availability of the first atomic bombs in February of 1945. But there was nothing inevitable about that timing. It was the net result of decisions and actions of many people, and their efforts could have yielded results at a different time. A case can be made, for example, that the US could have developed atomic bombs sometime in 1944.

The key scientific knowledge required for an atomic bomb project was available by the end of 1940. Scientists knew that U235 could be made to explode, and that it could be purified, albeit with a great effort. They also knew that plutonium could be made to explode, and that a nuclear pile could produce this new element. With the unfair advantage of hindsight, it would appear that the bomb project could have started earlier, say January 1941 instead of that June.[1] A Manhattan project begun earlier perhaps would have gotten off to a slower start, but still would likely have produced a bomb sooner than was the case. Why then did it take until 1945 to get the bomb? It was not so much the behaviour of radioactive elements but that of the human element.

In retrospect, there were several missed opportunities to get the bomb project initiated sooner than discussed in this book:

- Enrico Fermi's early 1939 meeting with War and Navy Department representatives. At that meeting the officials dismissed the physicists' predictions about the possibility of harnessing fission. It was 'not militarily useful' they said, taking little interest and no action.
- The three months, July to October 1939, that it took from Szilard's first drafting of a letter for Einstein to send to FDR, to the time when

---

1. See 'What Really Happened' (Jan 1942).

Alexander Sachs delivered and discussed it with FDR. Even in that pre-electronic age, preparing a letter need not have taken that long.
- The choice of Einstein to contact FDR. Genius he was, but as an academic, he evidently did not have the worldly sense of urgency the situation called for. In the post-war opinion of physicist Isidor Rabi, if Szilard had instead approached Nobel Laureate Ernest Lawrence of the University of California Berkeley, Lawrence would have ramrodded the bomb project to a far earlier start, and therefore an earlier culmination. Lawrence's ability in promoting 'Big Science' was such that, in Rabi's opinion, Lawrence would have gotten the bomb produced far earlier, perhaps in 1943 (Herken 2015).
- The six months Vannevar Bush spent on a series of National Academy Studies to review the findings of the Uranium Committee. As scientific data became available in 1940, Vannevar Bush could have used his position as science guru to persuade FDR to make a full-press effort toward a bomb. Instead he did the bureaucratic shuffle, commissioning three studies to review the subject. This deeply ingrained government pattern of calling for more data, more answers, before embarking on a course of action is an important safety mechanism to prevent ill-advised and wasteful efforts. But it can also delay important work, as here, with great cost in resources and lives. There are times when the government should move slowly and deliberately. This wasn't one of them.
- The initial leadership of the Manhattan Project by Colonel James Marshall (no relation to Army Chief of Staff George Marshall). The Army's Corps of Engineers was not put in charge of the project until June 1942, months after Roosevelt had greenlighted the building of a bomb. The first Corps manager, Colonel James Marshall, accomplished little in his several months before being replaced by Leslie Groves. Marshall was a solid, competent Corps of Engineers officer apparently more attuned to the decades-long construction projects of dams, canals and the like that were the Corps of Engineers' peacetime norm. James Marshall did not have the fire in his belly that Groves had.

Each of these missed opportunities 'delayed' the bomb's development by at least a few months. If any one or more had played out differently, it is entirely possible that the bomb could have been produced at least several months sooner than February 1945. Had the bomb instead been available in 1944, this would surely have altered the final stages of the war, and avoided many military and civilian deaths on both sides. One can only imagine how a 1944 end to the war would have shaped the post-war world differently.

But by the same token, several factors could have interacted to prevent the availability of the bomb in early 1945. As described earlier, a key decision by FDR was to appoint Walter Mendenhall to head a committee looking into atomic bomb development. FDR's advisor Edwin Watson had earlier identified Lyman Briggs as well qualified for this task, but Briggs' health problems at the time prompted Watson to suggest Mendenhall instead. Yet it is not hard to imagine the choice going to Briggs after all. Briggs' health problems were transitory (he continued doing research for many more years and died in 1968). Appointing Briggs to chair the Uranium Committee would likely have made a difference, given the timidity and cautiousness he demonstrated in his later work as Mendenhall's assistant. Recall British physicist Mark Oliphant's comment after he found that Briggs had hidden a key research report from the British in his office safe: Oliphant called Briggs an 'inarticulate and unimpressive man' (Rhodes 1986). How much time (and how many lives) might Briggs' bureaucratic cautiousness have cost if he had been in charge of the Uranium Committee?[2]

Beyond the Briggs/Mendenhall question, the appointment of a military man to head up the bomb development project was also critical. Leslie Groves was an excellent choice, but he was already involved in another major assignment: construction of the new office building now called the Pentagon. As noted earlier, Groves replaced Colonel James Marshall, himself a well-qualified, but far more deliberate, engineer. If James Marshall had remained at the helm of the Manhattan Project, or if Groves had not been put in charge in December 1941, the project would have been successful, but somewhat later.[3]

Thus, one or two different choices in the early 1940s could have yielded an atom bomb much later than was the case. How much later? Perhaps instead of February 1945, the bomb might have been ready a year later in early 1946. How would that have changed the end of the Second World War?

Certainly, the Nazis would have fallen even without the use of atomic bombs, probably before the end of 1945. As of January 1945 the planning assumption in Washington was that the European war would likely end in the late spring-early summer of that year, although there

---

2. See 'What Really Happened' (1940).
3. See 'What Really Happened' (Sep 1942).

was concern that some fighting could drag on if the Nazis took refuge in their reported Alpine Redoubt (*Time* 1945).

Once Germany fell, the Western Allies would have thrown everything they had at ending the Pacific War as quickly as possible. Some projections envisioned the war against Japan lasting until 1947 or 1948 (Huber), but Pentagon planners took it as an objective to end the Pacific War within one year of the end of the European War, before public support might begin to fade (Brower). The concern was that a prolonged war with an obviously 'defeated' Japan would prompt the public (especially the millions of families of servicemen) to demand it be ended, even with less than an unconditional surrender.

How would it have been if the bomb had become available a few months after the Nazis fell and while the war in the Pacific was still raging? Would President Truman have hesitated to use the bomb against non-Caucasian people, for fear of history labelling him a racist? That's doubtful. Much more likely is that he would have directed the bomb to be used as soon as it was available, to help stop the killing on all sides. Nevertheless, if the bomb had been used against the Japanese (but not against the Germans because it hadn't been available), there would likely have been lots of armchair academic second guessing.

And if the bombs hadn't been available before the end of 1945? Even among the few men who knew of the existence of the Manhattan Project, there was no assumption that it would produce a war-ending weapon whenever it arrived on the scene. So January 1945 saw discussion among the Joint Chiefs and the Pacific commanders about how best to end the Pacific War as quickly as possible, without counting on the atomic bomb. The Navy urged blockade and bombardment as the primary means to force Japan to quit. They called for landings on the coast of China, and capturing Formosa (the island we now call Taiwan). This would provide air bases supporting aerial bombardment of Japan. These actions would complement the naval encirclement that would starve Japan of resources (Skates). At that time, the Navy had the resources to put this plan into effect. It had over 6,700 ships, including over 100 aircraft carriers, 23 battleships, 72 cruisers, 377 destroyers and 232 subs (NHHC) .

In contrast, the Army's concept called for direct invasion of the southernmost Home Island, Kyushu, potentially toward the end of 1945. Once accomplished, this would provide air bases to support a subsequent invasion early in 1946 of the main island Honshu, where Tokyo is located (Skates). The Army too had the resources. Its strength of over 8 million personnel included 22 divisions (roughly 15,000 personnel each) already in the Pacific, with over 50 more divisions theoretically available for redeployment from Europe (Vannoy). The US Marine Corps also had about half a million troops, almost all already in the Pacific theatre.

The Army asserted that its invasion approach would more likely end the war sooner. The fact that the Army's plan would also kill a lot more US soldiers seemed to some to be not very important in the inter-service rivalry that still characterised the US military. A senior British Royal Air Force officer who worked with the US military throughout the war observed, 'The violence of inter-Service rivalry in the United States in those days had to be seen to be believed and was an appreciable handicap to their war effort' (Rearden).

If, as seems likely, the Army had prevailed in winning approval for its Operation Downfall, the 'direct invasion of Japan' approach to ending the Pacific War, the challenges would have been enormous. The presumptive commander would have been MacArthur. (He was senior to Nimitz, the other Pacific theatre commander, by one day [Rearden].) An invasion of Kyushu, followed by an invasion of Honshu, would have required staggering resources in terms of troops, and naval, air and materiel support. It would have been on a scale several-fold greater than the Normandy invasion (Allen and Polmar). This would have required moving hundreds of thousands of troops, mountains of equipment, and thousands of ships nearly halfway around the world from the European theatre to the Pacific – and how many of those troops and sailors would relish fighting in another war?

The planners of Overlord, the Normandy invasion, worked on the intricate plans for well over a year. The planners for an invasion of Kyushu would have had only about six months or so. Offsetting this would have been the greater experience in planning amphibious assaults. The staffs at Nimitz's and MacArthur's headquarters had planned and executed dozens of amphibious assaults. By late in the war, they had mastered the complex choreography of transferring combat and support troops, their vehicles, weapons, equipment and supplies, from ocean-going ships to flotillas of smaller vessels that would deposit them on hostile, barricade-filled shores, all the while using massive naval firepower from ships and aircraft to effectively support these landings. (Hornfischer).

But included in that successful island-hopping experience was also the late 1944 campaign on Leyte in the Philippines. There, MacArthur's staff had grossly misjudged the suitability of central Leyte for the construction of airfields. The delay in setting up usable airstrips in the soggy terrain allowed the Japanese to maintain air superiority for several weeks longer than had been anticipated (Weinberg). This intelligence blunder would not have been so embarrassing if MacArthur had not himself served for many years as military chief in the Philippines, and boasted about his expertise on the area (Weinberg). This is the same MacArthur who was so confident of his defence plan for the Philippines before 1941 that he declared: 'no Chancellery in the world will ever attempt to willfully attack the Philippines' (Gaerlan). When the Japanese did launch a surprise attack on his command in the Philippines with devastating results, it was of course the fault of his subordinates (Morris). (Making hyperbolically boastful statements contrary to fact, and assiduously taking credit but always shifting blame is a continuing American tradition.) Under MacArthur's self-admittedly infallible leadership, what other intelligence failures and mis-estimations would have occurred in a very large-scale, complex, long-distance invasion of an intensely hostile island nation about which far less was known?

In Europe, Eisenhower so worried about the Normandy invasion that he had prepared a message accepting full responsibility for its failure (Eisenhower 1944). But the preparations he had overseen were so thorough that the initial landings proceeded for the most part somewhat more successfully than expected. But would a MacArthur-led invasion of Kyushu, after a shorter planning window, have been as methodically planned? Would it have been a bold work of courageous genius that would add to MacArthur's glory? Or would it have been a catastrophe in which the inevitable toll of US casualties from enemy action was rivalled by the toll from inadequate, ego-driven planning? Then, how much would such an event have encouraged the Japanese military to fight to the last farmer?

So if the atom bombs had not yet been available toward the end of 1945, the US would have used some combination of blockade, bombardment and boots to end the Pacific War, and there is little doubt the US and its Western Allies would have succeeded. But Bolsheviks, too, would have played a role, perhaps welcomed, perhaps not, by the US. As discussed previously, Stalin's somewhat improvised February 1945 attack on Japanese forces was a much smaller version of what his military staff had planned to launch if they'd had a chance to send troops from Europe to the Far East. How far would the Soviets have gotten in their war against Japan? And what difference would it have made in the post-war world? One can only imagine.

What Really Happened

Oct 1939　　Alerted to the possibility of a uranium bomb by a letter from Szilard and Einstein, FDR appoints career bureaucrat Lyman Briggs to lead a Uranium Committee. Walter Mendenhall was available, but not selected. *(text p. 6)*

1940　　Lyman Briggs chairs the Uranium Committee, but both his medical condition and his bureaucratic style impair the pace of its work. By May 1941, physicist Phil Abelson quits as advisor to Briggs to 'get some work done' on U235 separation. Abelson notes that Briggs was 'a fine old gentlemen . . . but he was too old to lead a vigorous activity. This required . . . imaginative, and driving people. And Briggs was not of the . . . background to lead this thing' (Abelson). *(text p. 12) (Alternative speculation text p. 223)*

Dec 1940　　Seaborg and Segre discover plutonium, and find it to be fissionable. *(text p. 10)*

Sep 1941　　On a limited budget, in a basement at Columbia University, Enrico Fermi builds an experimental atomic pile to try to sustain a controlled nuclear chain reaction. Does not succeed. His next attempt, at the University of Chicago, succeeds. *(text p. 8)*

Nov 1941　　After ignoring British research on fission, and conducting its own, unhurried, research for two years, Briggs' Uranium Committee recommends initiating an atomic weapons programme. *(text p. 13)*

Jan 1942　　FDR approves atomic bomb project. *(text p. 221)*

Jun 1942	Army appoints Colonel James Marshall to manage construction for atomic bomb. Marshall proceeds deliberately. *(text p. 16)*
Jul 1942	As the roof is being put onto sections of the four-storey Pentagon, brass decides to add a fifth floor. Colonel Leslie Groves takes the change in his stride and delivers the heightened Pentagon on time. The additional space accommodates thousands of additional personnel, although the extra floor has less headroom than the other floors. And no outer windows. *(text p. 187)*
Sep 1942	Leslie Groves, just finished building the Pentagon at a hectic pace, takes over the bomb project ('The Manhattan Engineer District') with a similar approach. *(text p. 17) (Speculation text p. 223)*
2 Dec 1942	Team led by Enrico Fermi, formerly of Columbia University, initiates the first human-made self-sustaining nuclear chain reaction, at University of Chicago. *(text p. 13)*
Feb–Mar 1945 (2)	Battle of Iwo Jima. Kamikaze attacks sink one and damage two aircraft carriers out of more than two dozen involved in the battle. *(text p. 67)*
4 Feb 1945	Yalta Conference convenes. Roosevelt, Churchill and Stalin confer on plans for ending wars against Germany and against Japan, and on post-war dispositions. *(text p. 101)*
Early Feb 1945	In January 1945 the Soviet Red Army had made major westward advances across German-occupied Poland. Lead portions of Marshal Zhukov's front had reached the River Oder, less than 100 miles from Berlin. In early February, Stalin authorises Zhukov to get ready to push toward Berlin, first by bringing more divisions up to the line of the Oder River by 3 February, and then preparing to move against the Seelow Heights, the only topographic obstacle between his closest elements and Berlin. However, on 10 February, re-thinking his concern about Zhukov's exposed northern flank, Stalin put the drive to Berlin on hold, over the disagreement of several of his top commanders (Glantz 2001). He told Zhukov to redirect his efforts northward to protect his exposed flank. To Zhukov's south, during February, Konev's

	Front continued west, until it had advanced as far west as Zhukov. Soviet advance toward Berlin does not resume until 16 April (see below). *(text p. 44)*
13 Feb 1945	Allied firebombing raid on Dresden destroys 2.5 square miles of city and kills many thousands. Goebbels claimed at the time that the death toll had been 200,000; postwar analysis indicates the actual firebombing death toll was about 25,000, akin to that in other German cities (Eddy) *(text p. 52)*
13 Feb– 7 May 1945	The Allied firebombing of Dresden on 13–15 February 1945 killed 25,000. On 9 March, 100,000 people died in the firebombing of Tokyo. In other bombings in 1945, perhaps another 300,000–400,000 Germans and Japanese died. Also in that time, the Germans killed an estimated 250,000 people from concentration camps during death marches from Poland into Germany. In the Philippines, the February–March 1945 battle for Manila killed about 100,000 civilians. Not far away, in northern Vietnam, the hoarding of rice by Japanese occupiers contributed to the famine deaths of perhaps one million Vietnamese in the first half of 1945. Thus, war-related civilian deaths in the last several months before the defeat of the Nazis were likely far in excess of one million. After that, in the three months while the war in the Pacific continued, roughly another million civilians died in that theatre of war (Giangreco). *(text p. 85)*
15 Feb 1945	On 15 February German forces commanded by Himmler attacked south from Stargard seeking to penetrate Zhukov's northern flank, but the offensive stalled after three days. Nevertheless, this weak German attempt may have helped confirm Stalin in his determination to have Zhukov and Rokossovsky resolve the German threat from the north before resuming the push toward Berlin, which therefore did not begin again until mid-April. *(text p. 57)*
19 Feb 1945	US Marines land on Iwo Jima, completing its capture five weeks and tens of thousands of casualties later. Among other unexpected difficulties was the soft volcanic ash of the beaches (Wright). *(text p. 70)*

Feb–Mar 1945 (1)	Battle of Manila rages until early March, with 100,000 civilian deaths. *(text p. 70)*
Late Feb 1945	Wernher von Braun moves his rocket team and truckloads of documents and equipment from Peenemünde to Mittelwerk. *(Wartime text p. 80) (Post-war p. 164)*
9 March 1945	In Indochina, Japanese occupiers are alarmed by indications of imminent French resistance. Japanese suddenly depose the French colonial administrations; French officials and troops imprisoned in Vietnam, Laos and Cambodia. *(Surrender procedures text p. 141) (textbox p. 143)*
15 Mar 1945	Based on intelligence from Alsos teams, USAAF bombers destroy uranium refining facility at Oranienberg to deny it to Soviets after the war. *(Wartime text p. 78) (Post-war p. 160)*
Mar–Apr 1945	Peak of famine in Tonkin (northern Indochina), caused by combination of Japanese diversion of rice supply and flood damage to spring harvest. Deaths likely in excess of one million (Chandler et al.). *(text p. 142)*
Mid-Apr to mid May 1945	In Germany, US Army 104th Infantry Division enters Nordhausen, relieves prisoners, and ships many convoys of V-2 and V-1 missiles, jets and other advanced technology to the west, into designated US occupation zone. *(text p. 82)*
12 Apr 1945	President Roosevelt dies of cerebral haemorrhage about two months after returning from the mid-February Yalta Conference. *(text p. 104)*
16 Apr 1945	Months after initial units reached Oder River 40 miles east of Berlin, 2.2 million Soviet troops begin crossing river on a wide front to advance on Berlin. *(text p. xx)* At the Seelow Heights, elements of Zhukov's 1st Belorussian Front encounter unexpected difficulties crossing the marshy terrain just west of the Oder, taking four days to reach objectives planned for the first day. Stalin therefore authorises Konev, to the south, to also head for Berlin. *(text p. 56)*

30 Apr– 7 May 1945	Hitler commits suicide; 30 Apr. Surrounded by Soviet forces, Berlin falls after intense street fighting, 2 May. Admiral Dönitz briefly seeks better deal with Western Allies, but surrenders all forces to all Allies 7 May. *(text p. 65)*
2 May 1945	Wernher von Braun and several hundred rocket-builder colleagues surrender to US Army in southern Germany. Elsewhere, Soviets capture a few German rocket technicians and documents. *(Wartime text p. 80); (Post-war text p. 164)*
Early summer 1945	Pentagon's targeting committee develops list of A-bomb targets. Tokyo was not on the list, in part because so much of that city had already been destroyed by conventional and incendiary bombing. *(text p. 31)*
16 Jul 1945	Scientists successfully detonate first atomic explosion, a plutonium implosion device with a yield of 22,000 tons of TNT. Blast is seen and felt from Albuquerque NM to El Paso Texas (about 150 miles away). *(text p. 37)*
26 Jul 1945	Truman and Churchill issue Potsdam Declaration promising 'prompt and utter destruction.' if Japan did not surrender. *(text p. 39)*
6 Aug 1945	The first nuclear bomb used in warfare is dropped on Hiroshima, Japan. By that time, many other major Japanese cities had previously been extensively damaged by massive firebombing raids. The Great Yokohama Air Raid of 29 May, for example, had destroyed 42 per cent of the urban area (Tillman). *(text p. 50)*
9 Aug 1945	The second nuclear bomb used in warfare is dropped on Nagasaki by the B-29 *Bockscar*, after diverting from the mission's primary target, Kokura, due to meteorological conditions. *(text p. 61)*
9 Aug 1945 (2)	With about 1.6 million troops, USSR invades Japanese-held Manchuria, exactly three months after defeat of Nazis, as Stalin had promised. *(text p. 60)*
10 Aug 1945	With the Soviet Red Army making rapid gains against the Japanese in Manchuria, Pentagon planners Colonels Rusk and Bonesteel hastily recommend the 38th Parallel

	in Korea to demarcate the USSR vs US occupation areas. This provision becomes part of the surrender procedures given to Japan in General Order #1. *(text p 76.)*
15 Aug 1945 (1)	In Japan, after nine days of indecisive meetings of war councils and after trying to propose a conditional surrender, the Japanese government complies with the Emperor's directive to drop their conditions and surrender. Despite an attempted 'coup' by army officers, Emperor broadcasts surrender decree. *(Japan text p. 69) (Soviet text p. 96)*
15 Aug. 1945 (2)	News of Japan's surrender prompts street celebrations. In New York's Times Square, photographer Alfred Eisenstadt captures image of sailor kissing a woman in a white dress. *(text p. 94)*
17 Aug 1945	Truman approves General Order #1 on surrender procedures. US to take surrender on all Japanese Home Islands; US and USSR to split surrender responsibilities in northern vs southern Korea; GMD and UK to split responsibilities in northern vs southern Indochina. *(text p. 74)*
Aug 1945 – Oct 1949	In their August 1945 offensive, the Soviets swept through all of Manchuria and northern Korea, and also entered North China but did not reach Beijing or Tianjin. With US help, the GMD soon re-established control of North China, and made inroads into Soviet-occupied Manchuria. When the Soviets left, the Communist PLA was in control of most but not all of Manchuria, and were equipped with some Japanese materiel. From this substantial far-northern base Mao resumed his war against US-backed Chiang Kai-shek. Mao suffered reversals, but his greatest allies over the next several years were the arrogance, incompetence and corruption of Chiang and his commanders, which enabled the Communist armies to defeat the Nationalists, who fled to Taiwan. Mao declared the establishment of the People's Republic of China in October 1949. *(text p. 125)*
2 Sep 1945 Indo-china	In Indochina, as war ends, French colonial government has been deposed and imprisoned by Japanese since March. In the power vacuum, communist Viet Minh

	seek to take control. Ho Chi Minh persuades Emperor Bao Dai to abdicate. Ho Chi Minh issues Declaration of Independence for Democratic Republic of Vietnam. This is ignored by US. *(text p. 141) (text box p. 143)*
Sep 1945 (2)	Chinese Nationalist Guomindang (GMD) forces enter Tonkin and take surrender of Japanese forces in Indochina north of 16th Parallel; British arrive in Saigon and take surrender south of that line. French colonial officials seek to re-establish control of Indochina so that an Indochina Federation can be formed under French guidance and protection. *(text p. 142)*
9 Sep 1945	In accordance with General Order, #1 of 17 August 1945, US Army takes positions in the southern portion of Korea. US General Hodge takes surrender of Japanese forces there. Although not subject to US commander's authority, Soviets acted consistent with the General Order. On 26 August the Soviet Red Army had halted its advance from the north at the 38th Parallel line and had taken surrender of Japanese in the north. The 38th Parallel was intended only as a convenient dividing line for the Japanese surrender, not as a permanent political boundary. *(text p. 191)*
Oct–Dec 1945	Although the Potsdam Conference envisioned development of a single Korea-wide governance structure leading to full independence, the US promptly sets up a military government in their occupation zone in the south; they experience widespread dissatisfaction from the diverse Korean political factions. Soviets install a recently-returned Korean expatriate, Kim Il-sung, said to be a former guerrilla leader in Manchuria and a Soviet Red Army officer, as leader of the northern Communist Party in Pyongyang. Soviets also support Communist-led provisional governing structures in their northern occupation zone. *(text p.118)*
1946–8	As established at the December 1945 Four-Power Moscow Conference, a US-Soviet Joint Commission for Korea seeks to determine composition of a Korea-wide Provisional Government that would work under

	a Trusteeship toward independence. Disagreements between US and USSR, and among Korean political factions, preclude agreement on government composition. 'Temporary' occupation zones become North and South Korea. *(text p. 110)*
Jan 1946	US Air Forces award contract to Convair for the 'Hiroc', an intercontinental ballistic missile. *(text p. 184)*
Mar–May 1946 (1)	While Joint Commission for Korea meets in Seoul to try to set up a Korea-wide government, protests and street violence against occupation restrictions and conditions increase in Seoul and elsewhere. *(text p. 112)*
31 Aug 1946	John Hersey's *Hiroshima* published. *(text p. 53)*
Nov 1946–Feb 1947	In Korea, Soviets control northern part of country, and exploit inability of the Joint Commission to reconcile fractious Korean political groups and set up a Korea-wide provisional government. In November 1946 Soviets sponsor elections for local People's Committees throughout north; in February 1947 Soviets set up a Korean People's Assembly there. *(text p. 113)*
1946–2020	Beginning after the Second World War debut of the V-2, the US has continuously sought to develop a reliable anti-missile missile. Several nuclear-armed interceptor systems were deployed around US cities and missile sites in the 1950s and 1960s. The modern National Missile Defense system in Alaska relies instead on a kinetic intercept – hitting a bullet with a bullet. In tests, this system has proven to be effective, about half the time. *(text p. 184)*
Jul 1947	Curtis LeMay, touting his success in firebombing Japan during much of 1945, lobbies successfully for bomber funds. US Air Force, believing the Soviet Union lacked missile development capability, cancels the planned 'Hiroc' ballistic missile and redirects the funds to bombers and cruise missiles. *(text p. 185)*
Sep 1947	In Korea, US is unable to achieve acceptance among various Korean political parties for any provisional government scheme. Over Soviet objections, US refers the matter to the United Nations. *(text p. 115)*

Oct 1947	In northern Vietnam, during the French Operation Lea, Ho Chi Minh and other Viet Minh leaders escape capture by French paratroopers by a matter of minutes at Bac Kan, reportedly with help of a Viet Minh agent in French HQ. *(text p. 150)*
May 1948	UN Temporary Commission on Korea (UNTCOK), set up in 1947, supervises elections for a Constitutional Assembly. Due to Soviet opposition in the north, elections held only in southern Korea. Syngman Rhee's party wins most seats. Assembly elects Rhee President, who soon proclaims the (Korea-wide) independent Republic of Korea, but his Seoul-based government only holds sway in the south. *(text p. 116)*
Oct 1948	USSR recognises Kim Il-sung as Premier of Pyongyang-based Democratic People's Republic of Korea, holding sway in the north. *(text p. 119)*
Mar 1949	Bao Dai, former Emperor of Vietnam, agrees through the Paris Élyseé Accords to serve as Head of State of Vietnam, a free state within French Union. The State of Vietnam purports to include all of Vietnam but actually controls only southern Vietnam. *(text p. 146)*
29 Aug 1949	Soviet Union detonates their first atomic bomb. *(Oranienberg p. 160) (Rocket team p. 186)*
Sep 1949	Soviet Union tests R-2 missile, based on design of V-2 rockets captured from Germany. *(text p. 171)*
1 Oct 1949	In Beijing, Mao Zedong proclaims People's Republic of China. *(text p. 190)*
Jan 1950	President Truman orders development of hydrogen bomb. *(text p. 160)*
30 May 1950	In South Korea mid-term elections substantially weaken Syngman Rhee's legislative power. His support is dwindling due to the weak economy; the Soviets had cut trade between the rural south and the industrial north. Following these election results, Rhee is not certain he would be re-elected in 1952. The invasion by the north the next month changes that. *(text p. 120)*
May 1950	With Mao Zedong's victory in China, Communist establish a Chinese Military Advisory Group in Vietnam,

	sending weapons, materiel, and troops from People's Liberation Army to training and fight alongside Viet Minh (Westad). *(text p. 151)*
Mid-1950	From 1946 to mid-1950, Wernher von Braun and team tinker with captured V-2s on shoestring budget at Ft. Bliss, Texas in the desert south-west. After Army relocates them to Huntsville, Alabama, team is able to begin rocket work in earnest. *(text p. 175)*
Sep 1951 (1)	US Air Force approves development of Atlas ICBM, based on Hiroc design of 1947. *(text p. 185)*
1 Nov 1952	US detonates first hydrogen bomb. *(text p. 188)*
Early 1953	Atomic Energy Commission estimates that lighter-weight (2,000–3,000lb) nuclear weapons would be available in several years. USAF changes Atlas warhead requirement from 8,000lbs to 3,000lbs, prompting re-design and scaling down of Atlas missile for greater range and quicker launch preparation. *(text p. 189)*
5 Mar 1953	Soviet leader Josef Stalin dies. *(text p. 174)*
15 Mar 1953	Soviets test R-5 missile, capable of delivering nuclear warhead to European targets. *(text p. 172)*
27 Jul 1953	Korean War armistice freezes battle lines in place, almost exactly along the pre-war line, the 38th Parallel. Dividing line is demarcated by a 2.5-mile wide demilitarised zone. In August, South Korea and US sign a Mutual Defense Treaty under which US forces remain stationed in South Korea to the present day. Their presence near the DMZ is a tripwire. If the North were to invade again, they would immediately encounter American troops, and would be at war with the US. *(text p. 121)*
Mar 1954	In USSR, rocket engineer Sergei Korolev begins developing the R-7 missile to deliver 11,000lb warhead to intercontinental distances, or 3,000lbs to orbit. *(text p. 174)*
8 Sep 1954	Southeast Asia Treaty Organization (SEATO) established with US, France, the UK, Thailand, Australia, New Zealand, but none of the Indochinese states. *(see text p. 154)*

Nov 1954	CIA contracts Lockheed's 'Skunk Works' for an innovative high-altitude aircraft, the U-2 spy plane. First test flight occurs less than one year later, with first operational flight over Communist territory June 1956. In 1960, a U-2 piloted by Francis Gary Powers is shot down over the Soviet Union, leading to a major Cold War incident. In 1962, U-2 flights over Cuba detect placement of Soviets' nuclear-armed intermediate range missiles capable of reaching deep into US. *(text p. 206)*
9 Mar 1955	The popular TV show *Disneyland* airs *Man In Space* episode featuring descriptions of spaceflight and space stations narrated by Wernher von Braun as well as Willy Ley *(text p. 191)*
Jul 1955	President Eisenhower proposes 'Open Skies', allowing superpowers aerial reconnaissance of each other's military installations. Soviets summarily reject it. *(text p. 202)*
Jul 1955–1958	The USSR led the US to believe they had much greater long-range bomber capabilities than they actually possessed, as exemplified by a deceptively repetitive fly-by of their biggest bombers at a Moscow airshow. Partly in response to this, the US rapidly develops the U-2 spy plane. Photos by this aircraft showed by 1958 that there was a bomber gap – in favour of the US. The pattern repeated a few years later about a missile gap. Debate continues about the manipulation of concern about these gaps for partisan political purposes and out of interservice (Air Force/Navy/Army) rivalries. *(see text p. 210)*
Aug 1955	To supplement their limited ICBM capabilities, Soviets begin developing medium and intermediate range missiles; several types of these are deployed to Cuba in 1962. *(text p. 209)*
Oct 1955	In South Vietnam, Prime Minister Ngo Dinh Diem is dissatisfied with Geneva agreements; he wrests control of government and forecloses reunification elections. Vietnamese communists increase guerrilla warfare in South Vietnam. *(text p. 147)*

Nov 1955	Pending availability of Atlas ICBM, the Defense Department authorises Air Force to develop Thor, and the Army to develop Jupiter, as intermediate (1,500 miles) range missiles. In 1960, Jupiter missiles are deployed to Turkey, a Soviet neighbour, and to Italy, complementing Thor missiles deployed two years earlier to the UK. Jupiters play a role in the Cuban Missile Crisis of October 1963. *(text p. 210)*
Sep 1956	Army's von Braun rocket team launches Jupiter missile to altitude of 680 miles, and a range of 3,300 miles with a 30lb dummy satellite. Launch included an unfuelled fourth stage, as ordered by Pentagon due to interservice rivalry. With that extra fuel, satellite would have gone into orbit. *(See also 31 Jan 1958 below)*
Mar 1957	Engineer Korolev shows Soviet leader Khrushchev mock-up of R-7 ICBM. Khrushchev approves development of a satellite to be launched by the R-7. *(text p. 193)*
Jun 1957	US Air Force launches first test model Atlas ICBM with a range of just 600 miles. *(text p.203)*
Aug 1957	USSR launches R-7 capable of delivering 11,000lb warhead into US. *(text p. 195)*
4 Oct 1957	USSR uses R-7 to launch 184lb Sputnik I, first human-made earth satellite. Earlier that year, the CIA predicted a Soviet satellite launch by the end of 1957. President Eisenhower, while not surprised, was also dismissive of the Russian's feat, both publicly and privately. Underestimating Sputnik's propaganda value for Moscow, Ike had not wanted to let civilian space efforts 'distract' from ICBM development (Fortin). *(text p. 195)*
3 Dec 1957	US Navy uses Vanguard rocket to launch satellite – 4ft into the air before rocket explodes. *(text p. 203)*
31 Jan 1958	US launches 30lb Explorer, first American satellite, using a Jupiter rocket developed by Army's von Braun rocket team. *(See also Sep 1956, above)* *(text p. 203)*

1 May 1958	Soviets launch 3,000lb Sputnik 3, the much-delayed instrument package originally intended to be Russia's first satellite. *(text p. 206)*
Aug 1959	On thirteenth attempt, first successful orbital insertion and capsule recovery of Corona/Discoverer satellite, using a modified Thor intermediate range missile. *(text p. 205)*
Sep 1959	First US ICBM's (six Atlas-Ds) become operational, with 2200lb warheads at F.E. Warren AFB Wyoming *(text p. 211)*
26 Apr 1960	Repatriated to Korea in 1945 from the US, Syngman Rhee rapidly rises to Presidency of (South) Korea, with US backing. Rhee governs with a heavy hand through three Presidential terms that included invasion by North Korea. After the armistice, the South's economy recovered slowly, not helped by Rhee's repressive measures. Shortly after he was elected a fourth time, students and others vehemently protested the autocracy and corruption. Rhee was forced to resign and flee the country to exile in US. Although there is a Korean community in Annandale VA, there is no record he ever spent time there. *(text p. 121)*
1 May 1960	USSR shoots down U-2 spy plane piloted by Francis Gary Powers, leading to an international crisis that sets back efforts at arms limitation. *(text p. 206)*
23 Oct 1960	At Baikonur test range in USSR, a prototype of the USSR's first deployed ICBM, the R-16, explodes on the pad during a test, killing over 100 personnel including the head of the R-16 programme and senior missile designers. A faulty electrical switch was blamed (Chertok). *(text p. 171)*
12 Apr 1961	Soviet Cosmonaut Yuri Gagarin became first human in orbit, launched by an R-7 derivative. *(text p. 213)*
5 May 1961 and 20 Feb 1962	Alan Shepard becomes first American in space, launched atop a modified Redstone missile developed by the Army's von Braun rocket team. Despite the 'one step at a time' cover story, flight is 'sub-orbital' rather than fully orbital only because the US did not yet have a rocket powerful enough to get a manned capsule into

	orbit. On 20 February 1962, John Glenn becomes first American in orbit in a 3,000lb Mercury capsule launched on an Atlas-D rocket, a more powerful rocket than the Redstone. *(text p. 213)*
25 May 1961	Months after winning an extremely close race against Nixon, and days after the first American travelled in space, President Kennedy calls for US landing on the moon before end of the decade. *(text p. 219)*
Oct 1963	With a naval blockade and a military alert, President Kennedy demands the Soviets remove their newly installed intermediate range nuclear missiles from Cuba. At the end of the 13-day Cuban Missile Crisis, US secretly agrees to remove its (obsolete) Jupiter missiles from Turkey. *(text p. 210)*
Dec 1963	USAF announces the Manned Orbiting Laboratory programme to evaluate the 'military usefulness' of putting a human into space. Its classified aim was to put a crewed surveillance satellite into orbit to spy on the Soviet Union. The six-year programme never got into space (Howell). *(text p. 214)*

References Cited

Abelson, P. (1966), Interview with Stephan Groueff. Voices of the Manhattan Project. http://manhattanprojectvoices.org/oral-histories/philip-abelsons-interview-1966

Adelstein, J. (2015), 'New Evidence of Japan's Effort to Build Atom Bomb at the End of WWII', *Los Angeles Times* 5 August 2015, http://www.latimes.com/world/asia/la-fg-japan-bomb-20150805-story.html

Aldrich, R. (1998), 'British Intelligence and the Anglo-American "Special Relationship" during the Cold War', *Review of International Studies* 24 (3) 1998: 331–51.

Alexander, B. (1986), *Korea. The First War We Lost*. Hippocrene Books.

Allen, T. and Polmar, N. (1995), *Codename Downfall: The Secret Plan to Invade Japan - And Why Truman Dropped the Bomb*. Simon and Schuster.

Alsop, J. (1959), 'True Missile Gap Picture Belies Pentagon Response', *Eugene Register-Guard*, 13 October 1959.

Andrews, C. and Gordievsky, O. (1990), *KGB: The Inside Story of Its Foreign Operations from Lenin to Gorbachev*, Harper Collins.

Angell, J. (1953), *Historical Analysis of the 14–15 February 1945 Bombings of Dresden*. USAF Historical Division Research Studies Institute, Air University.

Anonymous (1945), *History of 509th Composite Group*. Air Force Historical Research Agency Archive.

Anonymous (1949), 'A Girdle Around The Earth', *Flight Magazine* 10 March 1949.

Anonymous (2004), *Konstantin E. Tsiolkovsky*. Aeronautics Learning Laboratory for Science Technology, and Research (ALLSTAR) Network.

Armstrong, C., and Post, J. (2004), *The North Korean Revolution, 1945–1950*. Cornell University Press.

Associated Press (1962), 'Soviet Fishing Port in Cuba is Scanned Closely', *Ellensburg Daily Record* 26 September 1962.

Atomic Heritage Foundation (undated), James Marshall Profile. http://www.atomicheritage.org/profile/james-marshall

Atomic Heritage Foundation (2014). *Project Silverplate*. https://www.atomicheritage.org/history/project-silverplate

Baker, B. (2015), 'What if the Kuomintang Had Won the Chinese Civil War?', *The Diplomat* 24 December 2015, https://thediplomat.com/2015/12/what-if-the-kuomingtang-had-won-the-chinese-civil-war/

Baldwin, S. (1932), *A Fear For The Future*. https://en.wikisource.org/wiki/A_Fear_For_The_Future

Battle, J. (undated), *Documents on the U.S. Atomic Energy Detection System [AEDS]*. National Security Archive Electronic Briefing Book No. 7 George Washington University. http://nsarchive.gwu.edu/NSAEBB/NSAEBB7/nsaebb7.htm

Baucom, D. (1992), *The Origins of SDI 1944-1983*. Modern War Studies, University Press of Kansas. http://www.amazon.com/The-Origins-1944-1983-Modern-Studies/dp/0700611002#reader_0700611002

Beevor, A. (2002), *Berlin: The Downfall 1945*. Penguin Books.

Behr, E. (1987), *The Last Emperor*. Bantam Books.

Bell, E. (2016), *Luna 2*. NASA Space Science Data Coordinated Archive. NASA. http://nssdc.gsfc.nasa.gov/nmc/spacecraftDisplay.do?id=1959-014A

Bender, J. (2014), 'The Astounding Devastation Of World War II', *Business Insider* 29 May 2014, https://www.businessinsider.com/percentage-of-countries-who-died-during-wwii-2014-5

Benford, G. (2017), *The Berlin Project*. Saga Press.

Benke, R. (1997), 'New Details Emerge About Japan's Wartime A-Bomb Program', *Los Angeles Times*, 1 June 1997.

Berger, A. (2014), 'What Lies Beneath: North Korea's Uranium Deposits', NK News.org. 28 August 2014, https://www.nknews.org/2014/08/what-lies-beneath-north-koreas-uranium-deposits/

Bernstein, J. (1992), 'The Farm Hall Transcripts: The German Scientists and the Bomb', *New York Review of Books*, 13 August 1992, https://www.nybooks.com/articles/1992/08/13/the-farm-hall-transcripts-the-german-scientists-an/

Bernstein, J. (2010), 'John von Neumann and Klaus Fuchs: an Unlikely Collaboration', *Physics in Perspective* 12 (1) 2010: 36–50.

Bernstein, R. (2014), *China 1945: Mao's Revolution and America's Fateful Choice*. Alfred Knopf.

Berthon, S. and Potts, J. (2007), *Warlords: An Extraordinary Re-creation of World War II Through the Eyes and Minds of Hitler, Churchill, Roosevelt, and Stalin*. Da Capo Press.

Bethe, H. (2000), 'The German Uranium Project', *Physics Today* 53: 7, 2000, https://doi.org/10.1063/1.1292473

Bischof, G. (2009), 'Allied Plans and Policies for the Occupation of Austria, 1938–1955', in: Steininger, Rolf et al. (2009), *Austria in the Twentieth Century*. Transaction Publishers.

Blair, Clay (2001), *Silent Victory: The U.S. Submarine War Against Japan*. Naval Institute Press

Blitz, M. (2016), 'When Kodak Accidentally Discovered A-Bomb Testing', *Popular Mechanics* http://www.popularmechanics.com/science/energy/a21382/how-kodak-accidentally-discovered-radioactive-fallout/

Borkin, J. (1978), *The Crime and Punishment of IG Farben*. Free Press.

Bowen, L. (1959), *Project Silverplate 1943–1946. The History of Air Force Participation in the Atomic Energy Program, 1943–1953 Vol. I*. US Air Force, Air University Historical Liaison Office.

Brians, P. (1987), *Nuclear Holocausts: Atomic War in Fiction, 1895-1984*. Kent State University Press, https://brians.wsu.edu/2016/11/16/nuclear-holocausts-atomic-war-in-fiction-3/

Broad, W. (2020), 'An Atomic Spy', *New York Times* 28 January 2020.

Brooks, C., Grimwood, J. and Swenson, L. (1979), *Chariots for Apollo: A History of Manned Lunar Spacecraft*. NASA Special Publication-4205.

Brooks, L. (1968), *Behind Japan's Surrender: The Secret Struggle to End an Empire*. Carpe Veritas Books.

Brower, C. (2012), *Defeating Japan. The Joint Chiefs of Staff and Strategy in the Pacific War, 1943-1945*. Palgrave MacMillan.

Brzezinski, M. (2007), *Red Moon Rising. Sputnik and the Hidden Rivalries That Ignited the Space Age*. Henry Holt and Co.

Burns, J. (1970), *Roosevelt: the Soldier of Freedom*. Harcourt Brace Jovanovich.

Burrows, W. (1999), *This New Ocean: The Story of the First Space Age*. Modern Library.

Burton, N. (undated), *1945 Treaty of Friendship and Alliance Between the Republic of China and the USSR*. http://www.chinaforeignrelations.net/

Buruma, I. (2013), *Year Zero. A History of 1945*. Penguin Press.

Bush, V. and Conant, J. (1944), *Memorandum from Vannevar Bush and James B. Conant, Office of Scientific Research and Development, to Secretary of War,*

Re: *Future International Handling of Subject of Atomic Bombs.* September 30, 1944. https://nsarchive2.gwu.edu//NSAEBB/NSAEBB162/1.pdf

Butow, R. (1954), *Japan's Decision to Surrender.* Stanford University Press.

Cadbury, D. (2007), *Space Race: The Epic Battle Between America and the Soviet Union for Dominion of Space.* Harper Perennial.

Campbell, V. (2004), 'How RAND Invented the Postwar World', *Invention and Technology* Summer 2004.

Cassidy, D. (2000), 'Copenhagen: A Historical Perspective', *Physics Today* July 2000.

Census Bureau (US) (1975), *Historical Statistics of the United States, Colonial Times to 1970.*

Center For Military History, US Army (1952), *Korea 1950.* https://history.army.mil/html/books/021/21-1/CMH_Pub_21-1.pdf

Chandler, D., Cribb, R. and Narangoa, L. (eds) (2016), *End Of Empire.* NIAS Press.

Chen, C., (2007), 'Heinz Guderian'. World War II Database. http://WWIIdb.com/person_bio.php?person_id=272

Chen, C., (2009), 'Battle of Lake Khasan'. World War II Database. http://WWIIdb.com/battle_spec.php?battle_id=232

Chen, C., (2014), 'Turkey'. World War II Database. https://WWIIdb.com/country/Turkey

Chertok, B. (2006), *Rockets and People, Volume 2: Creating a Rocket Industry.* https://history.nasa.gov/SP-4110/vol2.pdf

Chi, C. (2002), Taiwan Documents Project: 'The Surrender of Japanese Forces in China, Indochina, and Formosa' http://www.taiwandocuments.org/japansurrender.htm

'China, Soviet Union: Treaty of Friendship and Alliance. 1946.' *The American Journal of International Law*, 40(2), 51-63. www.jstor.org/stable/2213813.

Cima, R. (ed.) (1987), *Vietnam: A Country Study.* Library of Congress.

Cochran, T and Norris, R. (1995), *Making the Russian Bomb: From Stalin to Yeltsin.* Westview.

Coker, K. (2013), *U.S. Army Reserve Mobilization for the Korean War.* US Army Reserve Command.

Cole, H. (2005), *The Ardennes: Battle of the Bulge. US Army in World War II.* US Army Center For Military History, Office of the Chief of Military History, Department of the Army. https://history.army.mil/html/books/007/7-8-1/index.html

Connaughton, R., Pimlott, J. and Anderson, D. (1995), *The Battle for Manila.* Bloomsbury Publishing.

Conover, E. (2019), 'Uranium cube traced back to Nazis', *Science News* 22 June 2019.
Coster-Mullen, J. (2012), *Atom Bombs: The Top Secret Inside Story of Little Boy and Fat Man*. Waukesha, Wisconsin: OCLC 298514167.
Craven, W. and Cate, J. (undated), *The Army Air Forces in World War II Vol. 5: The Pacific: Matterhorn to Nagasaki June 1944 to August 1945*. Office of Air Force History, US Air Force, Washington DC.
Cully, G. (2020), *Operation Silverplate*. Cybermodeller Online https://www.cybermodeler.com/history/silverpl/silverpl.shtml?
Cummings, B. (1981), *The Origins of the Korean War, Liberation and the Emergence of Separate Regimes, 1945-1947*. Princeton University Press.
Cummings, B. (2005), *Korea's Place in the Sun: A Modern History*. W. W. Norton and Company.
Cummings, B. (2010), *38 Degrees of Separation: a Forgotten Occupation. The Korean War: a History*. Modern Library.
Curtis, H., (2010), *Rocket Vehicle Dynamics. Orbital Mechanics for Engineering Students*. Elsevier.
Dahl, P. (1999), *Heavy water and the wartime race for nuclear energy*. Institute of Physics Publishing.
Dallek, R. (2008), *Harry S. Truman*. Times Books.
Dash, M. (undated), 'Final straggler: the Japanese soldier who outlasted Hiroo Onoda'. https://mikedashhistory.com/2015/09/15/final-straggler-the-japanese-soldier-who-outlasted-hiroo-onoda/
Davenport, C. (2016), 'Why The Most Maligned Rocket In The World Is Also One Of The Most Reliable', *Washington Post*, 23 March 2016.
Davis, R. (2006), *Bombing the European Axis Powers. A Historical Digest of the Combined Bomber Offensive 1939–1945*. Air University Press.
Deane, H. (1999), *The Korean War, 1945–1953*. China Books and Periodicals.
De Bruhl, M. (2006), *Firestorm: Allied Airpower and the Destruction of Dresden*. Random House.
Department of the Army (US) (1966), *The Campaigns of MacArthur in the Pacific*. Reports of General MacArthur.
Department of Energy (US) (undated), *The Manhattan Project: An Interactive History*. Office of History and Heritage Resources. https://www.osti.gov/opennet/manhattan-project-history.
Department of Energy (US) (2002), *Hanford Site Historic District: History of the Plutonium Production Facilities, 1943–1990*. Battelle Press.
Department of State (US) (1945), *Inter-Allied Consultation Regarding Korea. Diplomatic Papers. Conferences at Malta and Yalta, 1945*. Office of the Historian.

Department of State (US) (1954), *Southeast Asia Treaty Organization (SEATO) 1954*. Office of the Historian.

Department of State (US) (1970), *The Conferences at Washington and Quebec, 1943. Foreign Relations of the United States*. Government Printing Office

Department of State (US) (undated), *(1). Allied Relations and Negotiations With Turkey* https://1997-2001.state.gov/www/regions/eur/rpt_9806_ng_turkey.pdf

Dobbs, M. (2012), *Six Months in 1945*. Alfred Knopf.

Dommen, A. (2001), *The Indochinese Experience of the French and the Americans*. Indiana University Press.

Dower, J. (1999), *Embracing Defeat: Japan in the Wake of World War II*. W. W. Norton and Company.

Drell, S. (1999), 'Physics and U.S. National Security', *Reviews of Modern Physics* 71:2.

Dryden, H., Pickering, W. and Tsien, H. (1946), *Guided Missiles and Pilotless Aircraft: A Report of the AAF Scientific Advisory Group*. HQ Air Materiel Command.

Dudden, A. (2006), *Japan's Colonization of Korea: Discourse and Power*. University of Hawaii Press.

Duffy, C. (1991), *Red Storm on the Reich: The Soviet March On Germany, 1945*. Da Capo Press.

Duiker, W. (2000), *Ho Chi Minh: A Life*. Hyperion Books.

Dykman, J. (undated), *The Soviet Experience in World War Two*. The Eisenhower Institute, Gettysburg College.

Ebrey, P. (1993), *Chinese Civilization: A Sourcebook (2nd ed.)*, Simon and Schuster.

Eckert, C., Lee, K., Lew, Y., Robinson, M. and Wagner, E. (1990), *Korea Old and New*. Ilchokak.

Eisenhower, D. (1944), 'Message Drafted by General Eisenhower in Case the D-Day Invasion Failed.' National Archives. https://www.archives.gov/education/lessons/d-day-message/

Eisenhower, D. (1948), *Crusade in Europe*. Doubleday.

Emme, E. (1965), *A History of Spaceflight*. Holt, Rinehart and Winston.

Everts, S. (2016), 'The Nazi Origins of Deadly Nerve Gases', *Chemical and Engineering News* 94:41, 17 October 2016. http://cen.acs.org/articles/94/i41/Nazi-origins-deadly-nerve-gases.html

Fall, B. (1961), *Street Without Joy The French Debacle in Indochina*. Stackpole.

Farmelo, G. (2013), *Churchill's Bomb: How The United States Overtook Britain in the First Nuclear Arms Race*. Basic Books.

Federation of American Scientists (undated), *Early Missile Developments*. http://www.fas.org/nuke/guide/usa/icbm/early.htm

Filippelli, R. (undated), *Vietnam Notebook*. www.parallelnarratives.com

Fine, L. and Remington, J. (1972), *The Corps of Engineers: Construction in the United States*. United States Army Center of Military History.

Fisk, R. (2000), 'Poison Gas from Germany', *The Independent* 30 December 2000.

Fogerty, R. (1953), *Biographical Data on Air Force General Officers, 1917–1952*. USAF Historical Division.

Fortin, J. (2017), 'Orbit of Sputnik Surprised Many, But American Spies Saw It Coming', *New York Times* 7 October 2017.

Frank, R. (1999), *Downfall: The End of the Imperial Japanese Empire*. Random House.

Frankland, N. and Webster, C. (1961), *The Strategic Air Offensive Against Germany, 1939–1945, Volume II: Endeavour, Part 4*. Her Majesty's Stationery Office.

Furlong, R. (2005), 'Hitler "tested small atom bomb"' BBC News 14 March 2005 http://news.bbc.co.uk/2/hi/europe/4348497.stm

Fussell, P. (1981), 'Thank God For The Atom Bomb', *New Republic* August 1981.

Gaddis, J. (1997), *We Know Now: Rethinking Cold War History*. Clarendon Press.

Gaddis, J. (2005), *The Cold War: A New History*. Penguin Press.

Gaerlan, C. (2012), 'General MacArthur and the Fall of Bataan and Corregidor', *Hyphen Magazine* May 2012. http://newamericamedia.org/2012/05/general-macarthur-and-the-fall-of-bataan-and-corregidor.php

Garlinski, J. (1978), *Hitler's Last Weapons: The Underground War against the V1 and V2*. Times Books.

Giangreco, D. (2009), *Hell To Pay. Operation Downfall and the Invasion of Japan*. Naval Institute Press.

Glantz, D. (1983), *August Storm: The Soviet 1945 Strategic Offensive in Manchuria*. Combat Studies Institute US Army Command and General Staff College Ft. Leavenworth Kansas.

Glantz, D. (2001), *The Soviet-German War 1941-1945: Myths and Realities*. Distinguished Lecture at the Strom Thurmond Institute of Government and Public Affairs Clemson University https://pl.b-ok2.org/book/2345339/b91a50

Goddard, E., and Pendray, G. (eds) (1970), *The Papers of Robert H. Goddard*. McGraw-Hill.

Goncharov, G. (1996). 'American and Soviet H-bomb Development Programmes: Historical Background', *Physics-Uspekhi* 39 (10) 1996.

Gordon, Y. and Rigmant, V. (2002), *Tupolev Tu-4: Soviet Superfortress*. Midland Counties Publications Ltd.

Gorin, P. (1997), 'Zenit: Corona's Soviet Counterpart', in McDonald. Robert A., *Corona Between the Sun and the Earth: the first NRO reconnaissance eye in space*. The American Society for Photogrammetry and Remote Sensing.

Goudsmit, S. (1947), *Alsos*. Henry Schuman Publishers.

Gray, P. (1995), 'Doomsdays. A Merciless War Comes To An Appalling End With the Use of Atomic Bombs and the Instant Incineration of Two Cities', *Time* 7 August 1995.

Green, C. and Lomask, M. (1970), *Vanguard: A History*. The NASA Historical Series NASA SP-4202 National Aeronautics and Space Administration.

Griffith, R. (1970), *The Politics of Fear: Joseph R. McCarthy and the Senate*. University of Massachusetts Press.

Groves, L. (1962), *Now It Can Be Told*. Harper.

Gunn, G. (2011), 'The Great Vietnamese Famine of 1944-45 Revisited', *The Asia-Pacific Journal: Japan Focus* January 2011.

Haboush, J. (2016), *The Great East Asian War and the Birth of the Korean Nation*. Columbia University Press.

Halliday, J. and Cumings, B. (1988), *Korea: The Unknown War* Viking Press.

Hammer, E. (1955), *The Struggle for Indochina 1940-1955: Vietnam and the French Experience*. Stanford University Press.

Han, S. (1975), 'Reviewed Work: Life of Kim Kyusik by Chong-sik Lee', *The Journal of Asian Studies* Vol. 34, No. 4, 1975.

Hardesty, V. and Eisman, G. (2007), *Epic Rivalry: The Inside Story of the Soviet and American Space Race*. National Geographic.

Harford, J. (1997), *Korolev. How One Man Masterminded The Soviet Drive To Beat America To The Moon*. John Wiley and Sons, Inc.

Hart-Landsberg, M. (1998). *Korea: Division, Reunification, & U.S. Foreign Policy*. Monthly Review Press.

Hasegawa, T. (2005), *Racing the Enemy: Stalin, Truman and the Surrender of Japan*. Belknap Harvard Press.

Hastings, M. (1988), *The Korean War*. Simon and Schuster.

Hastings, M. (2009a), 'Operation unthinkable: How Churchill wanted to recruit defeated Nazi troops and drive Russia out of Eastern Europe' *Daily Mail* 26 August 2009. http://www.dailymail.co.uk/debate/article-1209041/Operation-unthinkable-How-Churchill-wanted-recruit-defeated-Nazi-troops-drive-Russia-Eastern-Europe.html

Hastings, M. (2009b), *Retribution: The Battle For Japan 1944-45*. Alfred A Knopf.
Hastings, M. (2012), *Inferno, The World At War 1939-1945*. Alfred A Knopf.
Hatada, T., Smith. W. and Hazard, B. (1969), *A History of Korea*. ABC-Clio.
Hauben, J. (2018), *Is the UN Role in Korea 1945-1953 the Model Being Repeated Today?* https://www.researchgate.net/publication/326647556_Is_the_UN_Role_in_Korea_1945-1953_the_Model_Being_Repeated_Today_
Haulman, D. (1999), *The U.S. Army Air Forces in World War II. The Air Offensive Against Japan*. Air Force History and Museums Program. https://permanent.access.gpo.gov/lps51153/airforcehistory/usaaf/WWII/hittinghome/index.htm
Haynes, J.E., Klehr, H. and Vassiliev, A. (2009), *Spies: The Rise and Fall of the KGB in America*. Yale University Press.
Heisenberg, W. (1953), *Nuclear Physics*. Philosophical Library
Henthorn, W. (1971), *A History of Korea*. Free Press.
Heppenheimer, T. (1997), *Countdown A History of Spaceflight*. John Wiley and Sons.
Herken, G. (2002), *Brotherhood Of The Bomb*. Henry Holt and Co.
Herken, G. (2015), 'The Overlooked Physicist Behind the Bomb's Birth', *Washington Post* 12 July 2015.
Hess, G. (1972), 'Franklin Roosevelt and Indochina', *Journal of American History* 59:2, September 1972.
Hinsley, F., Thomas, E., Simkins, C. and Ransom, C. (1988), *British Intelligence in the Second World War, Volume 3, Part 2: Its Influence on Strategy and Operations*. HMSO.
Hodge, J. (1947), Telegram: Lieutenant General John R. Hodge to the Secretary of State, 20 August 1947. *Foreign Relations of the United States, The Far East, Volume VI* https://history.state.gov/historicaldocuments/frus1947v06/d577
Hoffman, D. (1993), *Operation Epsilon*. Translated at German History in Documents and Images. National Archives and Records Administration, RG 77, Entry 22, Box 164 (Farm Hall Transcripts). http://germanhistorydocs.ghi-dc.org/pdf/eng/English101.pdf
Holloway D. (1994), *Stalin and the Bomb*. Yale University Press.
Holt, T. (2005), *The Deceivers: Allied Military Deception in the Second World War*. Phoenix.
Hooten, E. (1991), *The Greatest Tumult: The Chinese Civil War*. Brassey Publishers.
Hornfischer, J. (2017), *The Fleet at Flood Tide*. Penguin Random House.
Hotton, R. and Davis, M. (2016), *Willow Run*. Arcadia Publishing.

Howell, E. (2017), *Manned Orbiting Laboratory Declassified: Inside a US Military Space Station*. Space.com. https://www.space.com/34661-manned-orbiting-laboratory-declassified-photos.html

Hoyt, E. (1986), *Japan's War: The Great Pacific Conflict*. Cooper Square Press.

Hsu, I. (1995), *The Rise of Modern China*. Oxford University Press.

Huber, D. (1988), *Pastel: Deception in the Invasion of Japan*. Combat Studies Institute U.S. Army Command and General Staff College, Ft. Leavenworth.

Huzel, D. (1981), *Peenemünde to Canaveral*. Greenwood Press.

Jager, S. (2013), *Brothers At War: The Unending Conflict in Korea*. W. W. Norton and Co.

Johansen, H. (1948), 'What Can Our Bombers Do Now?', *Popular Science* August 1948.

Jones, V. (1985), *Manhattan: The Army and the Atomic Bomb*. United States Army Center of Military History.

Karlsch, R and Walker, M. (2005), 'New light on Hitler's bomb'. physicsworld.com 1 June 2005.

Karnow, S. (1983), *Vietnam: A History*. Viking Press.

Kean, S. (2010), *The Disappearing Spoon: And Other True Tales of Madness, Love, and the History of the World from the Periodic Table of the Elements*. Little, Brown.

Keegan, J. (2006), *Atlas of World War II*. Harper Collins.

Kemp, R. (2012), 'The End of Manhattan: How the Gas Centrifuge Changed the Quest for Nuclear Weapons', *Technology and Culture* April 2012 DOI: 10.1353/tech.2012.0046. https://www.researchgate.net/publication/236766581

Kershaw, I. (2008), *Hitler: A Biography*. W. W. Norton and Company.

Kim, H. (1988), 'The American Military Government in South Korea 1945-1948', *Asian Perspective* Vol. 12, No. 1, 1988.

Kiyoko, K. (1983), 'Japanese Military Policy Towards French Indochina during the Second World War: The Road to the Meigo Sakusen (9 March 1945)', *Journal of Southeast Asian Studies* 14(2), 1983.

Knaack, M. (1988), *Encyclopedia of U.S. Air Force Aircraft and Missile Systems: Volume II: Post-World War II Bombers, 1945–1973*. Office of Air Force History.

Koeth, T. and Hiebert, M. (2019), 'Tracking the journey of a uranium cube', *Physics Today* May 2019, 72:5 https://doi.org/10.1063/PT.3.4202

Kurtz-Phelan, D. (2018), 'The Marshall Plan That Failed', *The Atlantic* 30 July 2018, https://www.theatlantic.com/international/archive/2018/07/failed-marshall-plan/564905/

La Feber, W (1975), 'Roosevelt, Churchill and Indochina 1942–1945', *American Historical Review* 80:5, December 1975.

Landa, E. and Nimmo, J. (2003), 'The Life and Scientific Contributions of Lyman J. Briggs', *Soil Science Society of America Journal* 67:3, 2003.

Lankov, A. (2002), *From Stalin to Kim Il Sung: The Formation of North Korea 1945–1960*. Rutgers University Press.

Lanouette, W. (2014), Interview with Atomic Heritage Foundation on Leo Szilard. http://manhattanprojectvoices.org/oral-histories/william-lanouettes-interview

Le Tissier, T. (1996), *Zhukov at the Oder: the decisive battle for Berlin*. Greenwood Publishing Group.

Lebow, N. (1988), 'Was Khrushchev Bluffing in Cuba?', *Bulletin of the Atomic Scientists* 44:3, April 1988.

Leckie, R. (1962), *Conflict: The History of the Korean War 1950-1953*. G. P. Putnam's Sons.

Lee, C. (1963), 'Politics in North Korea: Pre Korean War', *China Quarterly* 14, 1963.

Lee, H., Park, S. and Yoon, N. (2005), *New History of Korea*. Jimoondang.

Lee, J. (2006), *The Partition of Korea after World War II: A Global History*. Palgrave Macmillan.

Lee, K. (1984) (translated by E.W. Wagner and E.J. Shultz), *A New History of Korea* (rev. ed.). Ilchogak.

Lee, K. (1997), *Korea and East Asia: The Story of a Phoenix*. Greenwood Publishing Group.

Lehman, M (1963), *This High Man: The Life of Robert H. Goddard*. Farrar, Strauss, and Co.

Lenczowski, G. (1990), *American Presidents and the Middle East*. Duke University Press.

Leymarie, P. (1997), 'Deafening Silence on a Horrifying Repression', *Le Monde Diplomatique*. http://mondediplo.com/1997/03/02madagascar.

Lethbridge, C. (undated), 'Atlas Program Background', *Spaceline* http://www.spaceline.org/rocketsum/atlas-program-background.html

Levine, A. (1994), *The Missile and Space Race*. Praeger Publishers.

Ley, W. (1955), 'For Your Information', *Galaxy Magazine*, October 1955.

Lim, Y., Ng, Y., Tam, J. and Liu, D. (2016), 'Human Coronaviruses: A Review of Virus–Host Interactions', *Diseases*, 4(3) Sep 2016: 26.

Lipp, J. and Salter, R. (1954), *Project Feed Back Summary Report: Volume I*. RAND Corporation. http://www.rand.org/pubs/reports/R262z1.html.

Logevall F. (2012), *Embers of War: The Fall of an Empire and the Making of America's Vietnam*. Random House.

Lone, S. and McCormack, G. (1993), *Korea Since 1850*. Longman Cheshire Pty Limited.

Longmate, N. (1983), *The Bombers*. Hutchins and Company.

Low, M. (1990), 'Japan's secret war? "Instant" scientific manpower and Japan's World War II atomic bomb project', *Annals of Science* 47:4. http://www.tandfonline.com/doi/abs/10.1080/00033799000200281

Lunde, H. (2018), *Soviet Winter Offensive from the Vistula to Oder*, https://warfarehistorynetwork.com/daily/wwii/the-soviet-winter-offensive-from-the-vistula

Lynch, M. (2010), *Chinese Civil War 1945-1949* Essential Histories. Osprey Publishing.

Lyons, G. (1976), *The Russian Version of the Second World War*. Facts on File Publications.

MacDonald, C. (1973), *United States Army in World War II. European Theater of Operations. The Last Offensive*. Office of the Chief Of Military History Department of the Army.

Manvell, R. (2011), *Goering*. Skyhorse Press.

Marnham, P. (2013), 'Tracing the Congolese mine that fuelled Hiroshima', http://www.telegraph.co.uk/culture/10416945/Tracing-the-Congolese-mine-that-fuelled-Hiroshima.html

Marr, D. (1995), *Vietnam 1945: The Quest For Power*. University of California.

Matray, J. (1998), *Korea's Partition: Soviet American Pursuit of Reunification 1945-48*. Mt. Holyoke Press.

McCreedy, K. (1995), *Planning the Peace Operation Eclipse and the Occupation of Germany*. School of Advanced Military Studies US Army Command and General Staff College.

McMurran, M. (2008), *Achieving Accuracy: A Legacy of Computers and Missiles*. Xlibris Publishers.

McTaggart P. (2020), 'Death On The Oder: Soviet Advance on Berlin', https://warfarehistorynetwork.com/2020/01/17/death-on-the-oder-the-soviet-advance-on-berlin/

Metz, H. (1993), *Algeria: A Country Study*. Federal Research Division, Library of Congress.

Metz, H. (1994), *Madagascar: A Country Study*. Federal Research Division, Library of Congress.

Mieczkowski, Y. (2013), *Eisenhower's Sputnik Moment: The Race for Space and World Prestige*. Cornell University Press.

Miller, E. (2004), 'Vision, Power and Agency: The Ascent of Ngo Dinh Diem', *Journal of Southeast Asian Studies* 35 (3) 2004: 433–58.

Miller, E. (2013), *Misalliance: Ngo Dinh Diem, the United States, and the Fate of South Vietnam*. Harvard University Press.

Miller, R. and Wainstock, D. (2014), *Indochina and Vietnam: The Thirty-Five Year War 1940-1975*. Enigma Books.
Mindling, G. (2011), *U.S. Air Force Tactical Missiles*. Lulu.com.
Mitchell, D. (2016), *Bossart: America's Forgotten Rocket Scientist*. Mental Landscape.
Morison, S. (1956), *Leyte, June 1944 – January 1945. History of United States Naval Operations in World War II XII*. Little and Brown.
Morris, E. (2000), *Corregidor: The American Alamo of World War II*. First Cooper Square Press.
Mosely, P. (1950), 'The Occupation of Germany. New Light on How the Zones Were Drawn', *Foreign Affairs*, July 1950.
Motter, T. (1952), *The Persian Corridor and Aid to Russia*. Center of Military History United States Army. Government Printing Office.
Murashima, E. (2005), 'Opposing French colonialism: Thailand and the independence movements in Indo-China in the early 1940s', *South East Asia Research* 13 (3), 2005.
Myers, P. and Sengers, J. (1999), *Lyman James Briggs: A Biographical Memoir*. National Academy Press.
National Aeronautics and Space Administration (2015), *About Project Mercury*. https://www.nasa.gov/mission_pages/mercury/missions/program-toc.html
National Air and Space Museum (2002), *The Space Race*. Smithsonian Institution. https://airandspace.si.edu/exhibitions/space-race/online/index.htm
National Committee for the Investigation of the Truth About the Jeju April 3 Incident (2006), *Final Report*. Office of the Prime Minister, Republic of Korea. https://web.archive.org/web/20090224221736/http://www.jeju43.go.kr/english/sub05.html
Natural Resources Defense Council (undated), *Table of US Strategic Bomber Forces*. Archive of Nuclear Data, http://nrdc.org/nuclear/nudb/datab7.asp
Naval History and Heritage Command (2016), *Ship Force Levels*. http://www.history.navy.mil/research/histories/ship-histories/us-ship-force-levels.html#1945
Neufeld, J. (1990), *The Development of Ballistic Missiles in the US Air Force 1945-1960*. Office of Air Force History.
Neufeld, M. (1996), *The Rocket and the Reich*. Harvard University Press.
Neufeld, M. (2007), *Von Braun: Dreamer of Space, Engineer of War*. Alfred A. Knopf.
Nichols, K. (1987), *The Road to Trinity*. William Morrow and Co.

Nohlen, D., Grotz, F. and Hartmann, C. (2001), *Elections in Asia: A data handbook, Volume II*. Oxford University Press Scholarship Online.

Nolan, T. (1975), *Walter Curran Mendenhall: A Biographical Memoir*. National Academy Press.

Norris, R. (2002), *Racing For The Bomb: General Leslie Groves, the Manhattan Project's Indispensable Man*. Steerforth Press.

NuclearFiles.org (undated), Transcript of phone conversation, Senator Harry Truman and Secretary of War Henry Stimson, 17 June 1943

Oberdorfer, D. (1997), *The Two Koreas: A Contemporary History*. Addison Wesley.

Oder, F., Fitzpatrick, J. and Worthman, P. (1987/declassified 2010), *The Corona Story*. National Reconnaissance Office, http://www.nro.gov/foia/docs/foia-corona-story.pdf

Office of Management and Budget (2016), *Fiscal Year 2017 Historical Tables*. Budget of the US Government.

ORAU (1999), *Fiesta Ware*. Oak Ridge Associated Universities https://www.orau.org/ptp/collection/consumer%20products/fiesta.htm

Ordway, F. and Mitchell, R. (1979), *The Rocket Team*. Apogee Books Space Series 36. Thomas Y. Crowell.

Ortmeyer, P. and Makhijani, A. (1997), 'Let Them Drink Milk'. Institute for Energy and Environmental Research 2009 originally published as 'Worse Than We Knew'. *The Bulletin of the Atomic Scientists*, November/December 1997. http://www.ieer.org/latest/iodnart.html

Padfield, P. (1990), *Himmler: Reichsführer-SS*. Henry Holt and Company.

Parker, J. (2006), 'U.S. Presidential Election, 1952', in *Encyclopedia of American Political Parties and Elections* 2006. American History Online. Facts On File http://www.fofweb.com/activelink2.asp?

Parsch, A. (2005), *RTV-A-2. Encyclopedia Astronautica Directory of U. S. Military Rockets and Missiles*, http://www.astronautix.com/lvs/atlasd.htm

Parsch, A. (undated), *Atlas D. Encyclopedia Astronautica Directory of U.S. Military Rockets and Missiles*, http://www.astronautix.com/lvs/atlasd.htm

Patti, A. (1980), *Why Viet Nam? Prelude to America's Albatross*. University of California Press.

Pedlow, G. and Welzenbach, D. (1992), *The Central Intelligence Agency and Overhead Reconnaissance: The U-2 and Oxcart Programs, 1954–1974*. Central Intelligence Agency.

Perez, L. (2013), *Japan at War: An Encyclopedia*. ABC-CLIO

Pratt, K. and Rutt, R. (1999), *Korea: A Historical and Cultural Dictionary*. Routledge.

REFERENCES CITED

Powers, T. (1994), *Heisenberg's War: The Secret History of the German Bomb*. Knopf.

Rafalko, F. (undated), *Counter-Intelligence in WWII*. Federation of American Scientists.

Ragheb, M. (2016) *Japanese Nuclear Weapons Program*. http://mragheb.com/NPRE%20402%20ME%20405%20Nuclear%20Power%20Engineering/Japanese%20Nuclear%20Weapons%20Program.pdf

RAND (1946), *Preliminary Design of an Experimental World-Circling Spaceship*. Report SM 11827. RAND Corporation. http://www.rand.org/pubs/special_memoranda/SM11827.html

RAND (undated), *Preliminary Design of an Experimental World-Circling Spaceship* website Abstract. http://www.rand.org/pubs/special_memoranda/SM11827.html

Read, A. and Fisher, D. (1992), *The Fall of Berlin*. W. W. Norton and Co.

Rearden, S. (2012), *Council of War: A History of the Joint Chiefs of Staff 1942–1991*. Joint History Office Joint Chiefs of Staff.

Reed, C. (2011), 'Correcting a Persistent Manhattan Project Statistical Error', *Bulletin of the American Physical Society*. http://meetings.aps.org/Meeting/APR11/Session/H13.4

Reed, T. (2004), *At the Abyss: An Insider's History of the Cold War*. Ballantine Books.

Reiman, V. (1979), *Joseph Goebbels: The Man Who Created Hitler*. Sphere Publishers.

Resis, A. (1978), 'The Churchill-Stalin Secret "Percentages" Agreement on the Balkans, Moscow, October 1944', *American Historical Review* 83:2, 1978.

Reuters (2010), 'France to return South Korea royal books on lease', 12 November 2010.

Reynolds, D. (2007), *Summits: Six Meetings That Shaped The Twentieth Century*. Basic Books,

Rhodes, R. (1986), *The Making of the Atomic Bomb*. Simon and Schuster.

Rhodes, R. (1995), *Dark Sun: The Making of the Hydrogen Bomb*. Touchstone Simon and Schuster.

Rosen, M. (1955), *The Viking Rocket Story*. Harper and Brothers.

Rosenberg, Z. (2014), 'Did a B-24 Really Shoot Down a V-2 Rocket in 1944?', *Air and Space Smithsonian Magazine*, 20 October 2014. http://www.airspacemag.com/daily-planet/did-B-24-really-shoot-down-V-2-rocket-1944-180953085/

Ross, I. (1968), *The Loneliest Campaign: The Truman Victory of 1948*. New American Library.

Ross, R. (1987), *Cambodia: A Country Study*. Federal Research Division, Library of Congress.

Roulo, C. (2019), '10 Things You Probably Didn't Know About the Pentagon', *DoD News*, 3 January 2019. https://www.defense.gov/Explore/Features/story/Article/1650913/10-things-you-probably-didnt-know-about-the-pentagon/

Ruffner, K. (1995), *Corona: America's First Satellite Program*. CIA Center For The Study of Intelligence.

Rusk, D. (1991), *As I Saw It: Memoirs of a Secretary of State*. Tauris.

Russell, R. (1997), *Project HULA: Secret Soviet-American Cooperation In The War Against Japan*. Naval Historical Center, Department of Navy.

Ryan, A. and Keeley, G. (2017), 'Sputnik and US Intelligence: the Warning Record', *Studies in Intelligence* 61:3, 2017.

Ryan, C. (ed.) (1952–4), Man Will Conquer Space Soon series. *Collier's Weekly Magazine*. http://www.unz.org/Pub/Colliers-1952mar22

Savada, A. (1994), *Laos: A Country Study*. Library of Congress.

Savada, A. and Shaw, W. (eds) (1990), *South Korea: A Country Study*. Library of Congress.

Sayre, J. (1945), *Persian Gulf Command*. Random House.

Schnabel, J. (1992), *United States Army in the Korean War: Policy and Direction, the First Year*. U.S. Army Center Of Military History.

Scott, D. and Leonov, A. (2006), *Two Sides of the Moon: Our Story of the Cold War Space Race*. St. Martin's Griffin.

Seth, M. (2010), *A Concise History of Modern Korea*. Rowman and Littlefield.

Severo, R. (1993), 'John Hersey, Author of "Hiroshima," Is Dead at 78', *The New York Times* 25 March 1993.

Sheehan, N. (2009), *A Fiery Peace in a Cold War: Bernard Schriever and the Ultimate Weapon*. Random House.

Shirer, W. (1983), *The Rise and Fall of the Third Reich*. Fawcett Crest.

Siddiqi, A. (2003), *Sputnik and the Soviet Space Challenge*. The University of Florida Press.

Skates, J. (1994), *The Invasion of Japan: Alternative to the Bomb*. University of South Carolina Press.

Smith, R. (1978), 'The Japanese Period in Indochina and the Coup of 9 March 1945', *Journal of Southeast Asian Studies* 9 (2), 1978.

Snell, D. (1946), 'Japan Developed Atom Bomb; Russia Grabbed Scientists', *Atlanta Constitution*. http://www.reformation.org/atlanta-constitution.html

Spector, R. (2007), *In the Ruins of Empire*. Random House.

Speer, Albert (1970), *Inside the Third Reich*. Macmillan.

Steininger, R. (2008), *Austria, Germany, and the Cold War*. Berghahn Books.

Stoker, J. (2004), *Britain and Ballistic Missile Defence 1941-2002*. Frank Cass/ Taylor and Francis Publishers. https://books.google.com/books

Strategic Bombing Survey (US) (1945), *Summary Report (European War)*. US Government Printing Office.

Stueck, W. (1984), *The Wedemeyer Mission: American Politics and Foreign Policy during the Cold War*. University of Georgia Press.

Stueck, W. and Yi, B. (2010), 'An Alliance Forged in Blood: The American Occupation of Korea, the Korean War, and the US–South Korean Alliance', *The Journal of Strategic Studies* 33:2, 2010.

Sublette, C. (2007), *Section 8.0 The First Nuclear Weapons; Little Boy*. Nuclear Weapons Archive. http://nuclearweaponarchive.org/Nwfaq/Nfaq8.html

Sublette, C. (2019), *Nuclear Weapons Frequently Asked Questions* https://nuclearweaponarchive.org/Library/Implsion.html

Svitak, A. (2012), 'Falcon 9 RUD?', *Aviation Week*, 26 November 2012.

Swenson, L., Grimwood, J. and Alexander, C. (1989), *This New Ocean: A History of Project Mercury*. NASA Special Publication-4201.

Szanton, A. (1992), *The Recollections of Eugene P. Wigner*. Plenum.

Takaki, R. (1995), *Hiroshima: Why America Dropped The Atomic Bomb*. Little Brown and Co.

Tanaka, Y. (1988), 'Poison Gas, the Story Japan Would Like to Forget', *Bulletin of the Atomic Scientists* October 1988.

Target Committee (1945), *Minutes of the second meeting of the Target CommitteeLos Alamos*. U.S. National Archives.

Taylor, F. (2005), *Dresden: Tuesday 13 February 1945*. Bloomsbury.

Taylor, J. (2009), *The Generalissimo: Chiang Kai-shek and the Struggle for Modern China*. Harvard University Press.

Teague, P. (2016), *The Soviet Invasion of Manchuria and the Kwangtung Army*. Paul S. Teague, Amazon Digital Services.

Tertitskiy, F. (2019), 'How an obscure Red Army unit became the cradle of the North Korean elite', NK News, 4 February 2019. https://www.nknews.org/2019/02/how-an-obscure-red-army-unit-became-the-cradle-of-the-north-korean-elite/

Thomas, E. (2013), *Ike's Bluff: President Eisenhower's Secret Battle to Save the World*. Back Bay Books.

Thomas, M. (1997), 'Free France, the British Government and the Future of French Indo-China, 1940-45', *Journal of Southeast Asian Studies* Vol. 28, No. 1, 1997.

Tibbets, P. (1998), *Return of the Enola Gay*. Enola Gay Remembered Inc.

Tillman, B. (2010), *Whirlwind: The Air War Against Japan 1942–1945*. Simon & Schuster.

Time (12 February 1945) 'World Battlefronts: Battle of Germany: The Man Who Can't Surrender'.

Toland, J. (1970), *The Rising Sun: The Decline and Fall of the Japanese Empire, 1936–1945*. Random House.

Tomasevich, Jozo (2001), *War and Revolution in Yugoslavia, 1941–1945: Occupation and Collaboration*. Stanford University Press.

Truman, H. (1945), 'Statement by the President Announcing the Use of the A-Bomb at Hiroshima', Presidential Speeches, Harry S. Truman. Miller Center, University of Virginia.

Truman, H. (1949), 'Statement Announcing the First Soviet A-Bomb', Atomic Archive. http://www.atomicarchive.com/Docs/Hydrogen/SovietAB.shtml

Truman, H. (1950), 'Truman Orders Hellbomb Made', *NY Daily News* 1 February 1950.

Truman, H. (1956), *Memoirs Vol 2. Years of Trial and Hope 1946-1953*. Doubleday.

Umeda, S. (2016), *National Parliaments: South Korea*. Library of Congress. https://www.loc.gov/law/help/national-parliaments/southkorea.php)

United Nations (1947), General Assembly Resolution 112 (II), 17 Nov 1947, http://daccess-ods.un.org/access.nsf/Get?OpenAgent&DS=A/RES/112(II)&Lang=E&Area=RESOLUTION

United Nations Environment Program (2011), *Viet Nam Assessment Report on Climate Change*. Institute of Strategy and Policy on Natural Resources and Environment.

United States Air Force (2017), *Soviet Union Impounds and Copies B-29*. National Museum of the USAF.

Van Allen, J. (undated), *James Van Allen Papers*. The University of Iowa Archives. https://www.lib.uiowa.edu/scua/bai/halas.htm

Van de Ven, H. (2018), *China At War: Triumph and Tragedy*. Harvard University Press.

Vannoy, A. (2014), 'Expanding the Size of the U.S. Military in World War II'. Warfare History Network. https://warfarehistorynetwork.com/2017/06/26/expanding-the-size-of-the-u-s-military-in-world-war-ii

Vogel, S. (2008), *The Pentagon: A History*. Random House.

Von Karman, T. (1945), *Where We Stand: Volume II of Toward New Horizons*. Report to the US Army Air Forces Chief Of Staff. http://www.governmentattic.org/TwardNewHorizons.html

Wade, M. (ed.) (undated), 'A9/A10', *Encyclopedia Astronautica*. http://astronautix.com/

Wade, M. (ed.) (undated), 'R-9', *Encyclopedia Astronautica*. http://astronautix.com/

Walden, G. (2000), 'Mittelwerk V-1/V-2 Rocket Factory, Nordhausen Dora Concentration Camp Site'. Third Reich in Ruins, http://www.thirdreichruins.com/

Walker, M. (1993), *German National Socialism and the Quest for Nuclear Power 1939–1949*. Cambridge.

War Department Equipment Board (1946), *Report to the Chief of Staff of the Army*. Washington DC. http://cgsc.cdmhost.com/cdm/ref/collection/p4013coll11/id/782

Ward, R. (2005), *Doctor Space: The Life of Werner von Braun*. Naval Institute Press.

Wasser, A. (2005), 'LBJ's Space Race: what we didn't know then (part 1)', *The Space Review*.com 20 June 2005. http://www.thespacereview.com/article/396/1

Wattendorf, F., Tsien, H. and Duwez, P. (1946), *Aircraft Power Plants. A Report of the AAF Scientific Advisory Group*. HQ Air Materiel Command.

Weinberg, G. (1994), *The World At Arms*. Cambridge University Press.

Wells, K. (1990), *New God, New Nation: Protestants and Self-reconstruction Nationalism in Korea, 1896-1937*. University of Hawaii Press.

Welsh, W. (2012), 'Masterful Defense at Seelow Heights'. Warfare History Network. https://warfarehistorynetwork.com/2017/06/08/masterful-defense-at-seelow-heights/

Welzenbach, D. (undated), *Observation Balloons and Reconnaissance Satellites*. http://www.foia.cia.gov/sites/default/files/document_conversions/89801/DOC_0000253109.pdf

Werth, A. (1966), *De Gaulle: A Political Biography*. Simon and Schuster.

Westad O. (2003), *Decisive Encounters: The Chinese Civil War 1946-1950*. Stanford University Press.

Wilcox, R. (1985), *Japan's Secret War: Japan's Race Against Time to Build Its Own Atomic Bomb*. Morrow Publishing.

Wilson, M. (2000), *Korean Government Publications: An Introductory Guide*. Scarecrow Press.

Woodhouse, C. (2002), *The Struggle for Greece 1941–1949*. Hurst & Company.

Wright, D. (2004), *Iwo Jima 1945: The Marines Raise the Flag On Mount Suribachi*. Osprey Publishing Ltd.

Zak, A. (undated (1)), 'Sputnik 1, Sputnik 2, Sputnik 3', RussianSpaceWeb.com. http://www.russianspaceweb.com/sputnik.html

Zak, A. (undated (2)) 'R-3 Rocket Family', RussianSpaceWeb. http://www.russianspaceweb.com/r3.html

Ziemke, E. (1972), *Battle for Berlin: End of the Third Reich*. Ballantine Books.

Zoellner, T. (2009), *Uranium. War, Energy and the Rock That Shaped The World*. Viking Penguin.

Zubok, V. (2007), *A Failed Empire: The Soviet Union in the Cold War From Stalin to Gorbachev*. UNC Press.

Zwetsloot, J. (April 2010), 'Sixty Years On, Memory of Korea's Gandhi Still Alive'. Korea.net

Index

A
Abelson, Phil, 227
Acheson, Dean, 120
Albania, 105
Aleutian Islands of Alaska, 77, 120
Algeria, 152
Alpine Redoubt, 84, 224
Alsos Project, 20–1, 23, 78, 80, 83, 230
Anami, General Korechika, 51, 61, 65
Annam, 131, 134, 145, 146
Antwerp, 79
Apollo Program, 216, 218
Arnold, General Henry 'Hap', 34, 36, 177–8, 179–180, 184, 185
Asada, Professor, 50–1, 65
Australia, 115
Austria, 84, 121
Avro Lancaster, 34, 36
Axis alliance, 14, 17, 24–7, 33, 38, 39, 53, 55, 63, 85, 93, 97, 101, 102, 105, 128, 158
Ayme, General Georges, 131, 134
Azerbaijan, People's Republic of, 97

B
Baldwin, Hanson W., 215
Baldwin, Stanley, 182
Bao Dai, 145, 146, 147, 149, 150, 151, 233, 235
Barnstable, Paul, 47
Baruch, Bernard, 198
Bavaria, 84
Beijing, 88, 123, 125, 232, 235
Belgian Congo, 17, 20, 21, 78
Belgium, 20, 38
Beria, Lavrenti, 159, 193
Berlin, 4, 26, 31, 43, 51, 52, 55, 56, 57, 62–3, 80, 84, 96, 163, 228, 231
Bidault, Premier George, 152
Bohr, Niels, 5, 19, 41
Bonesteel, Colonel Charles H., 231
Bossart, Karel, 180
Boushey, General Homer, 197, 198
Bradbury, Norris, 158
Braun, Wernher von, 79, 163–9, 170–4, 176–7, 188, 194, 231, 236, 238
Briggs, Lyman, 6, 12, 36, 223, 227
British Royal Navy, 74
Broz, Marshal Josip, 102, 119
Bulganin, Nikolai, 193
Bulgaria, 101, 102, 105, 114
Bulge, Battle of the , 21, 38, 39, 84
Burma, 39, 73
Bush, Dr Vannevar, 13–19, 158, 183, 184, 188, 222

Byelorussian Soviet Socialist
    Republic, 115

C
Cambodia, 131, 154
Cam Ranh Bay, 153
Canada, 3, 28, 115
Cao Bang, 129, 150
Cao Dai, 142, 146, 150, 151
Capitol Hill, 115, 185, 211
Carpenter, Scott, 217
Changchun, 87
Chelomei, Vladimir, 169, 173
Cherwell, Lord, 181
Chiang Kai-shek, 46, 87, 73,
    123, 124, 125, 132, 142, 144,
    152, 232
China, 29, 58, 73, 75, 90, 104,
    107, 113, 120, 124, 224
    People's Liberation Army, 46
Chinese Civil War, 149
Cho Man-sik, 110, 116
Churchill, Winston, 3, 10, 28, 37,
    38, 39, 45, 75, 93, 101, 102, 103,
    104, 105, 181, 228, 231
Clarke, Arthur C., 191
Cochin China, 127–8, 131, 134,
    142, 146
Compton, Arthur, 13, 19, 24
Conant, James, 17, 158
Cooper, Gordon, 217
Czechoslovakia, 41, 5, 102, 105, 161

D
Dalien, 45, 46, 87
Decoux, Admiral Jean, 74, 127–8,
    129, 131–5, 140, 141, 144
De Gaulle, General Charles, 74,
    128, 129, 143, 144, 146
Diebner, Kurt, 39, 41, 42
Diem, Ngo Dinh, 146, 152, 237

Dong, Pham Van, 139, 150
Dönitz, Admiral Karl, 231
Dornberger, General Walter, 179
Downfall, Operation, 225
Dresden, 47, 51, 52, 53, 57, 63, 83–6,
    85, 159, 229
Dryden, Hugh, 178
Duân, Lê, 151
Dunning, John, 7–8

E
Eatherly, Captain Claude, 61
Eden, Anthony, 143
Einstein, Albert, 5–6, 53, 222, 227
Eisenhower, Dwight, 39, 38, 47, 53,
    57, 64, 82, 84, 190, 191, 192,
    197, 199, 201–03, 205, 206,
    210–12, 216–19, 226, 237, 238
Eisenstadt, Alfred, 232
El Salvador, 115
Elbe, River, 96, 102
Élyseé Accords, 146, 235

F
FDR. *See* Roosevelt, Franklin D.
Feedback Project, 200, 201
Fermi, Enrico, 5, 7, 8, 12, 13, 19, 39,
    158, 221, 227, 228
Finland, 96, 114, 121
Flyorov, Georgy, 159
Formosa, 125, 224
Forrestal, James, 23
France, 4, 20, 26, 75, 95, 96, 105,
    115, 119, 128, 132, 141–7, 152,
    153, 163, 172, 217, 236
Frankfurt, 27
French Indochina Army, 74, 75,
    129, 130, 142, 150
French Navy, 74, 152
Frisch, Otto, 4, 10
Fuchs, Klaus, 158, 171

262

Fulbright, J. William, 197
Furman, Robert, 187

G
Gagarin, Yuri, 213, 239
Gemini Project, 216, 217
German Army, 79, 163
Germany, 1, 3, 19, 20, 25, 27–9, 36, 39, 40–2, 44, 47, 49, 63–5, 73, 76–8, 84, 93, 96, 115, 175, 230
Ghangzhou, 125
Giacobbi, Paul, 143
Giap, Vo Nguyen, 138, 139, 142, 143, 149, 150–1
Glenn, John, 217, 240
Glushko, Valentin, 163, 169, 173
Goddard, Robert, 166, 173, 175–7, 179, 181
Goebbels, Joseph, 52, 84, 229
Göring, Hermann, 64
Great Britain, 4, 10, 14, 19, 29, 47, 53, 102, 153, 182
Great Kanto Earthquake (1923), 61
Great Yokohama Air Raid, 231
Greece, 102, 105, 119
Greenglass, David, 171
Grissom, Gus, 217
Grottrup, Helmut, 165
Groves, Brigadier General Leslie, 16–19, 23, 26, 28, 34, 36, 159, 184, 187, 222, 223, 228
Guadalcanal, 107
Guam, 66, 94
Guderian, Heinz, 64

H
Hahn, Otto, 4, 39, 41
Haiphong, 134, 153
Hall, Theodore, 171
Hamburg, 27, 51
Hanford, 19, 20, 35

Hanover, 63, 83–6
Harbin, 60, 68, 71, 87–9
Hawaii, 107
Heilongjiang Province (Manchuria), 73–4, 87, 123
Heisenberg, Werner, 19, 41
Hersey, John, 53, 234
Himmler, Heinrich, 56, 64, 229
Hirohito, 50, 68, 69
Hiroshima, 231
Hitler, Adolf, 5, 20, 23, 25, 31, 38, 39, 44, 62, 63, 64, 95, 101, 175, 231
Ho Chi Minh, 123, 137, 138, 149, 150, 151, 233, 235
Hoa Hao, 142, 145, 146, 150
Hoan, Nguyen Ton, 147, 152
Hodge, Lieutenant General John, 90–1, 107–14, 117, 233
Holy Roman Empire, 114
Honshu, 29, 224, 225
Hula Project, 77
Hungary, 101, 102, 105

I
Idaho, 90
India, 115
Indiana, 30
Indochina, 31, 127, 137, 143, 152–4, 230, 232, 233
Itagaki, General Seishirō, 91
Italy, 4, 95, 114
Iwabuchi, Rear Admiral Sanji, 70–1
Iwo Jima, 47, 53, 66, 67, 70, 228, 229

J
Japan, 1, 4, 23, 25, 27, 28, 36, 42, 44, 48, 52, 66, 69, 70–1, 73, 74, 115, 127, 131, 232

263

Japanese Army, 30–1, 141
   First Area Army, 87
   Fourth Independent Army, 87
   Kempeitai, 68
   Kwantung Army, 31, 45, 58, 71, 76, 87, 88
   Seventeenth Area Army, 91
   Third Area Army, 60
Japanese Home Islands, 73, 75
Japanese Navy, 62
Jilin, 87, 88, 89
Jodl, Generaloberst Alfred, 64
Johnson, Kelly, 206
Johnson, Lyndon, 198, 211, 218
Jones, Dr Reginald V., 182

K
Karman, Dr Theodore von, 175–9, 181
Kempeitai, 68
Kennedy, John F., 211, 218, 240
Khrushchev, Nikita, 193, 194, 209, 238
Kido, Marquis Koichi, 65, 68, 69
Kiel, 27
Kim Il-sung, 117, 118, 119, 233, 235
Kim Ku, 108, 110, 117
King, Ernest, 23
Kistiakowsky, George, 14, 218, 219
Kobe, 61, 63, 66
Koiso, Kuniaki, 61
Kokura, 68, 133, 231
Kolankiewicz, Lieutenant Colonel Leon, 81–2
Konev, Marshal Ivan, 43, 44, 55, 56, 57, 63, 64, 165, 228–9, 230
Konoe, Prince Fumimaro, 66
Korea, 23, 52, 73, 75–7, 89–90, 107, 108, 114, 116, 118, 120
Korean Liberation Army, 117, 119
Korean Provisional Government, 108
Korolev, Lieutenant Colonel Sergei, 163–9, 171–4, 193–6, 204, 209, 213, 217, 236, 238
Kurchatov, Igor, 159
Kuril Islands, 45, 60
Kwantung Army, 31, 58, 71, 76, 87, 88
Kyushu, Home Island, 29, 224, 225, 226
Kyusik, Kim, 110, 111, 115, 116

L
Lake Khasan and Nomonhan, Battles of, 45–46
Laos, 127, 131, 134, 146, 154, 230
Larson, Arthur, 199
Lawrence, Ernest, 12, 222
Lawson, Art, 218
Lea, Operation, 150, 151
Leipzig, 64, 83–6, 159
LeMay, General Curtis, 47, 178, 187–8, 211, 234
Leyte Gulf, Battle of, 62, 67, 225
Ley, Willy, 191
Liaoning, 87, 88, 89
Libya, 97
Liège, 79
Lindemann, Professor Friedrich, 10, 181
London, 11, 25, 68, 79, 114, 165
Los Alamos, 18, 26, 34, 36, 37, 157–9, 171
Lyuh Woon-hyung. *See* Yo Un-hyung

M
MacArthur, General Douglas, 47, 53, 70, 73, 75, 89, 91, 111, 112, 141, 225, 226

Madagascar, 142, 152
Mahabad, Kurdish Republic of, 97
Malaya, 73, 127
Malecki, Master Sergeant Thad, 91
Malenkov, Georgi, 168, 193
Manchukuo, 58, 73, 76, 88, 88, 123
Manchuria, 30, 31, 45, 60, 61, 66, 71, 73, 77, 87, 88, 118, 123, 124, 125, 135, 231, 232
Manhattan Project, 157–8, 179, 184, 221, 223
Manila, Battle of, 53, 70, 229, 230
Mao Zedong, 46, 88, 118, 123, 125, 149, 151, 152, 153, 232, 235
Marshall, General George, 23, 36, 50, 78, 124, 183, 222
Marshall, Colonel James, 15–17, 222, 223, 228
Meitner, Lise, 4
Mekong River, 151–2
Mendenhall, Walter, 6–7, 9, 10–13, 14, 16, 18, 36, 223, 227
Menjiang, 58, 88, 123
Molotov, Vyacheslav, 112, 193
Montana, 90
Mordant, General Eugene, 128, 131, 133, 134, 143
Morocco, 152
Moscow (Kremlin), 26, 43, 55, 68, 94, 114, 119, 192, 197, 206, 209, 210
Mountbatten, Admiral Lord Louis, 73, 74, 84, 140
Mukden, 87
Mus, Paul, 138
Mutual Defense Treaty, 236

N
Nagasaki, 231
Nanjing, 114, 125

NATO, 153
Neumann, John von, 188–9
Neuss, 27
Nier, Alfred, 7–8
Nimitz, Fleet Admiral Chester William, 47, 73, 225
Nishina, Dr Yoshio, 23
Nixon, Richard, 191, 192, 198, 217, 218, 240
Nobel, Alfred, 157, 158
Nordhausen, 20, 80–2, 164–5, 171, 230
North Dakota, 90, 91, 212
Norway, 11, 19, 41

O
Oak Ridge, Tennessee, 16–20, 34
Oberth, Hermann, 173
Oder River, 43, 55, 56, 79, 202–06, 228, 230
Oikawa, Admiral Koshirō, 61
Okido, Lieutenant General, 68
Oliphant, Mark, 12, 223
Oppenheimer, J. Robert, 18, 19, 26, 36, 158
Oranienburg, 78, 159–60

P
Pak Hon-yong, 110, 116, 118
Paris Élyseé Accords, 235
Pas de Calais, 172
Pauley, Edwin, 112
Pearl Harbor, 17, 23, 45, 197, 198, 218
Peenemünde, 32, 57, 79–80, 163, 164, 170, 175, 177, 179, 230
Peierls, Rudolph, 10
Peking. *See* Beijing
Pentagon, the, 34, 73–5, 76, 115, 160, 223, 224, 228, 231
Philippines, 39, 70, 73, 74, 89, 229

Poland, 6, 43, 55, 96, 101, 102
Port Arthur, 45, 46, 87, 124
Portugal, 24
Powers, Francis Gary, 217, 237, 239
Projekt Amerika, 163, 166
Purnell, Rear Admiral William, 28

R
Rabi, Isidor Isaac, 12, 158, 222
Red Army, 6, 26, 44, 57, 60, 62, 63, 66, 77, 78, 82, 87–9, 94, 96, 104, 105, 117, 118, 123, 159, 163–5, 193, 228, 231, 233
  1st Belorussian Front, 43, 57, 80, 230
  1st Ukrainian Front, 43, 165, 228–9
  2nd Belorussian Front, 43, 56, 95
  Konev's Front, 56
  Soviet Fronts, 55
  Zhukov's Front, 56
Redoubt, Alpine, 224
Red River, 144, 146, 151–2
Rhee, Syngman, 110, 111, 113, 114, 116, 117–21, 235, 239
Ribbentrop, Joachim von, 64
Rokossovsky, General Konstantin, 43, 56, 229
Romania, 96, 101, 102, 105, 114
Roosevelt, Franklin D., 6, 7, 10, 19, 28, 37, 38, 39, 45, 47, 50, 69, 89, 101, 103, 104, 105, 143, 144, 222, 228, 230
Rusk, Lieutenant Colonel Dean, 76, 231
Russia, 75, 76, 97, 101, 102, 105, 107, 169, 209, 210, 212, 218.
  See also Soviet Union
Ryukyu Islands, 120

S
Sachs, Alexander, 6, 222
Saigon, 132, 134, 140, 142, 233
Saipan, 58, 66
Sakhalin Island, 45, 60, 61, 74
Sandburg, Carl, 198
Sandys, Duncan, 182
Schirra, Wally, 217
Schriever, Colonel Bernard, 180, 181, 183–6, 188, 192
Seaborg, Glenn, 9, 13, 24, 227
Seborer, Oscar, 171
Segre, Emilio, 9, 227
Shenyang. *See* Mukden
Shepard, Alan, 213, 214, 216, 217, 239
Shigemitsu, Mamoru, 61, 62, 66
Siberia, 45, 89
Sicily, 95
Sidey, Hugh, 218
Siegfried Line, 53
Sino-Soviet Treaty of Friendship, 89
Siping, 87
Solomon Islands, 107
Soong Tse-vung, 46
Southeast Asian Treaty Organization (SEATO), 153, 236
Soviet Army. *See* Red Army
Soviet Far Eastern Command, 57–8, 87
Soviet-Japanese Neutrality Pact, 45
Soviet Maritime Provinces, 108
Soviet Navy, 76–7
Soviet Union, 26, 27, 38, 44, 75, 77, 88, 89, 94–6, 101, 103, 104, 105, 110, 117, 160–2, 168–9, 171, 183, 193, 195–6, 202, 203, 209, 211, 212, 231–2, 234–9.
  See also Russia

Spaatz, Tooey, 178, 185
Speer, Albert, 62, 179
Stalin, Josef, 26, 37–9, 43–6, 55–7, 75, 77, 95–7, 101–03, 103–05, 159, 165, 168, 169, 183, 193, 226, 228, 236
Stargard, Battle of, 57
Staten Island, 17
Stettinius, Edward, 23
Stevenson, Adlai, 190
Stilwell, General Joseph, 183
Stimson, Henry, 16–17, 23, 37
Strasbourg, 20, 21
Strassman, Fritz, 4
Sulzberger, Cyrus L., 198
Sweden, 4, 24, 105
Sweeney, Major Charles, 47
Switzerland, 24
Symington, Stuart, 211
Syria, 115
Szilard, Leo, 5, 7, 8, 12, 13, 19, 24, 39, 221, 222, 227

T
Tadayoshi, Professor Hikosaka, 23
Tagaki, Admiral Sokichi, 62
Taylor, Captain Rick, 63
Teller, Edward, 5, 158, 197
Terauchi, Marshal Hisaichi, 140
Thailand, 154, 236
Thuringia, 77, 81
Tianjin, 123, 232
Tibbets, Colonel Paul, 33, 47
Tinian Island, 47, 61, 66, 161
Tito. *See* Broz, Marshal Josip
Tizard, Henry, 10–11
Tokyo, 26, 29–1, 43, 50, 52, 68–72, 231
Tonkin, 127, 129, 131, 133–5, 138–41, 143, 144, 149, 230
Toronto, 34

Truman, Harry, 37–8, 104, 144, 160, 186, 190, 231, 232
Tsiolkovsky, Konstantin, 163, 173, 176
Tsuchihashi, Japanese General, 133, 135, 140, 143
Tyuratam, 168, 174, 194, 195, 196

U
Ukraine, 193
Umezu, General Yoshijirō, 61
Union of Soviet Socialist Republics (USSR). *See* Soviet Union
United Kingdom (UK), 3–4, 10, 14, 19, 25, 28, 29, 38, 47, 53, 75, 102, 113, 153, 182
United States (US), 3–4, 17, 18, 25, 38, 75, 113,158, 232, 234
United States Army, 30, 81
  7th Infantry Division, 90
  21st Infantry Division, 70
  31st Infantry Division, 90
  41st Infantry Division, 90, 91
  104th Infantry Division, 81, 230
  509th Composite Group, 33, 36, 42, 47
  Seventh Army, 53
  First US Army Group, 172
  Military Government, 109
  US Army Air Forces (USAAF), 29, 33, 69, 214
USSR. *See* Soviet Union
US Navy, 77, 202

V
Valluy, General Jean-Étienne, 150–1
Vasilevsky, Marshal Aleksandr M., 57, 60, 71, 74, 87, 96
Verne, Jules, 215

Viet Minh, 127, 129, 137–9, 140, 141, 142, 145, 147, 149–52, 232–3, 235, 236
Vietnam, 123, 125, 127, 128, 137, 138, 142, 143, 145–7, 150–4, 229, 230
Vietnam Liberation Army, 139, 142
Vladivostok, 76, 94, 162

W
Washington DC, 30, 68, 83, 90, 93, 104, 114, 115
Watson, Edwin, 6, 223
Wells, Herbert G., 3, 5, 37
Westwall. *See* Siegfried Line
Whipple, Fred, 191
White, Thomas, 191
Wigner, Eugene, 7, 12
Wilson, Charles, 191

Y
Yamada, General Otozō, 71, 74
Yangel, Mikhail, 169, 173
Yo Un-hyung, 108, 110, 111, 116
Yokohama, 47, 50, 53, 61, 66, 71
Yonai, Admiral Mitsumasa, 51, 61, 62, 67
Yugoslavia, 102, 119

Z
Zhukov, Marshal Georgi, 43, 44, 55, 56, 63, 64, 80, 96, 228–30